建筑立场系列丛书 No.89

学习中的城市
Learning Cities

[日]SANAA建筑事务所 等 | 编
贾子光 段梦桃 | 译

大连理工大学出版社

004 巴克里希纳·多西的建筑——记2018年普利兹克建筑奖获得者
社会多元化：建筑的力量、社区的重要性与优势 _ Gihan Karunaratne

步行观景桥

008 伊丽莎白码头大桥 _ Arup Associates

014 姆扎巴大桥 _ Carinthia University of Applied Sciences + buildCollective

020 斯蒂尔树顶人行道 _ Glenn Howells Architects

026 维特雷P.E.M交通枢纽项目 _ TETRARC

034 萨默斯镇大桥 _ Moxon Architects + Arup

作为抵抗方式的文化

040 作为抵抗方式的文化 _ Francesco Zuddas

046 塞尔维亚卡伊莎文化中心 _ Guillermo Vázquez Consuegra

062 贝尔工作室——加拿大国家音乐中心所在地 _ Allied Works Architecture

074 鹤冈文化大厅 _ SANAA

088 麦德林现代美术馆扩建 _ 51-1 + Ctrl G

102 塞纳河音乐厅 _ Shigeru Ban Architects Europe

学习中的城市
把学校当成社交游乐场

124 学习中的城市——充满社交生活的学校 _ Andrea Giannotti

132 哥本哈根诺德海文国际学校 _ C.F. Møller Architects

148 丹麦Frederiksbjerg学校 _ Henning Larsen Architects + GPP Arkitekter

166 斯蒂芬·佩尔斯基金会高中 _ Chadwickdryerclarke Studio

182 斯科夫巴肯学校 _ CEBRA

196 马德里德语学校 _ Grüntuch Ernst Architects

206 建筑师索引

004 The Architecture of Balkrishna Doshi, the 2018 Pritzker Architecture Prize Laureate
— Social Pluralism: Power of architecture, Pride of place and strength in community _ Gihan Karunaratne

Bridge for Walk and View

008 Elizabeth Quay Bridge _ Arup Associates

014 Mzamba Bridge _ Carinthia University of Applied Sciences + buildCollective

020 Stihl Treetop Walkway _ Glenn Howells Architects

026 P.E.M Vitré _ TETRARC

034 Somers Town Bridge _ Moxon Architects + Arup

Culture for Resistance

040 Culture for Resistance _ Francesco Zuddas

046 CaixaForum Sevilla _ Guillermo Vázquez Consuegra

062 Studio Bell, Home of the National Music Center _ Allied Works Architecture

074 Tsuruoka Cultural Hall _ SANAA

088 Medellin Modern Art Museum Expansion _ 51-1 + Ctrl G

102 La Seine Music Hall _ Shigeru Ban Architects Europe

Learning Cities
The School as Social Playground

124 Learning Cities - Schools of Social Life _ Andrea Giannotti

132 Copenhagen International School Nordhavn _ C.F. Møller Architects

148 Frederiksbjerg School _ Henning Larsen Architects + GPP Arkitekter

166 Stephen Perse Foundation Senior School _ Chadwickdryerclarke Studio

182 Skovbakken School _ CEBRA

196 German School Madrid _ Grüntuch Ernst Architects

206 Index

巴克里希纳·多西的建筑——记2018年普利兹克建筑奖获得者
社会多元化：建筑的力量、社区的重要性与优势
The Architecture of Balkrishna Doshi, the 2018 Pritzker Architecture Prize Laureate
Social Pluralism: Power of architecture, Pride of place and Strength in community

巴克里希纳·多西
Balkrishna Doshi

受到印度迷宫一般的传统城市和寺庙的启发，班加罗尔的印度管理学院以连锁建筑、庭院和走廊的形式排布。它还提供了各式各样的空间来抵御炎热的气候，并通过半开放的走廊和花园将绿意带入室内。
"如今，位于班加罗尔的这座校园的显著特征在于集功能、运动、自然于一身，同时还能举办不同类型的活动。"

Inspired by traditional maze-like Indian cities and temples, Indian Institute of Management Bangalore is organized as interlocking buildings, courts and galleries. It also provides a variety of spaces protected from the hot climate, and infuses greenery through semi-open corridors and gardens.
"The integration of functions, movements, nature and the choices to accommodate diverse activities simultaneously are the distinguishing features of the campus at Bangalore today."

印度管理学院，班加罗尔，印度，1977—1992年。
多西手绘的平面图
Indian Institute of Management, Bangalore, India, 1977–1992.
Hand drawn plan by Doshi

内院
Internal courtyards

　　巴克里希纳·多西是印度独立后最具影响力的建筑师之一，他获得了2018年普利兹克奖。

　　多西是第一位获得这一建筑学领域最高殊荣的印度本土建筑师。在嘉奖评语中，普利兹克评审团称赞他的作品体现了"一种深切的责任感，以及通过高品质、真正的建筑为其国家和人民做出贡献的愿望"。

　　最初，普利兹克奖被授予在美国、欧洲和日本建造标志性建筑的著名"建筑大师"。然而，该奖项因为在建筑领域弘扬名人文化而一再遭到批评，有人批判这个奖项把重点放在了个人的创造性工作上，而不去关注建筑设计、城市主义和城市建设的集体性质。这就提出了以下两个问题：一是建筑在城市环境中的作用，二是建筑如何以一种更积极、更有意义的方式去促进城市的发展。

　　此外，该奖项的重点已经转移到反映行业内更丰富的多样性方面，鼓励采用更广泛的设计方法，也考虑到了更广泛的建筑范围。

　　在过去的十年中，该奖项颁发给了那些努力提高生活品质、通过建筑设计回应社会和经济需求，从而对建筑环境产生了更大积极影响的建筑师。之前的获奖者包括王澍（2012年）、坂茂（2014年）和亚历杭德罗·阿拉维纳（2016年），以及其他解决了诸多复杂社会问题的建筑师，他们积极参与社会建筑设计，通过采用当地可持续材料和方法来解决贫困问题和全球住房危机。

Balkrishna Doshi, one of the most influential architects of post-independence India, was named the 2018 laureate Pritzker Prize winner.

Doshi is the first native Indian architect to receive architecture's highest accolade. In its citation, the Pritzker jury praised his work as embodying "a deep sense of responsibility and a desire to contribute to his country and its people through high-quality, authentic architecture".

Initially, the Pritzker was awarded to renowned "star architects", who had built iconic buildings in the US, Europe and Japan. However, the prize was repeatedly criticized for endorsing the celebrity culture in architecture by concentrating on the creative work of individuals rather than focusing on the collective nature of building design, urbanism and city building. This raised questions regarding the role of architecture in the urban context and how architecture might contribute to the development of the city in a more positive and meaningful way.

Additionally, the focus of the prize has shifted to reflect a greater diversity within the profession, as well as celebrating a broader design approach, and bringing into consideration a wider range of buildings.

In the past decade, the prize has been given to architects who have made a greater positive impact on the built environment by attempting to improve the quality of life and respond to social and economic needs through architecture. Past examples of laureates include Wang Shu (2012), Shigeru Ban (2014), and Alejandro Aravena (2016), and other architects who have addressed complex issues through socially engaging architecture and by adopting local and sustainable materials and methods to address poverty, and global housing crisis.

Wang Shu (2012) stated that in China a high percentage of traditional buildings were demolished to make way for new

"我想开发一种系统,让班加罗尔的印度管理学院的建筑消失,再让这些建筑之间的空间来主导这个地方的体验……我认为最重要的事情是'拉莎'(古典印度曲风的新世纪组合),这是一种微妙的空间体验,使空间令人难忘。它扩展了人类的联想,也丰富了想象。"

"I wanted to develop a system by which the buildings at IIM Bangalore disappear and spaces in between them dominate the experience of the place…I thought the most important things [were] the rasa, which is the subtle experience of the space that makes the space memorable. It extends the associations and enriches the imagination."

阿兰若低成本住宅通过房屋、庭院和迷宫一般的内部通道系统容纳了8万多人。该社区由6500多套住宅组成,分为六个部分,每个部分都有一系列住房户型可供选择,从普通的一居室到宽敞的大户型住宅,适合不同收入人群。
"它们不是房子,而是幸福社区生活的家园。这才是最重要的。""看来我应该发个誓,一辈子记住这句话:为最低阶层的人提供合适的住所。"

Aranya Low Cost Housing accommodates over 80,000 individuals through a system of houses, courtyards and a labyrinth of internal pathways.
The community is comprised of over 6,500 residences, amongst six sectors – each of which features a range of housing options, from modest one-room units to spacious houses, to accommodate a range of incomes.
"They are not houses but homes where a happy community lives. That is what finally matters."
"It seems I should take an oath and remember it for my lifetime: to provide the lowest class with the proper dwelling."

从半开放的走廊向图书馆望去
View towards the library from a semi-open corridor

阿兰若低成本住宅的内部通道
Aranya Low Cost Housing internal pathway

街道缩影,作者:多西
Perspective of a street as a miniature, by Doshi

王澍(2012年)指出,在中国,为了给新建筑让路,大部分传统建筑都被拆除了。他试图通过使用被拆除建筑的材料建造新建筑来抵消这种影响。这是由记忆重新组成的建筑,突出了它的工艺、时间感和在此居住过的人们的记忆。¹ 坂茂为灾难和冲突频发地区设计的临时难民收容所和亚历杭德罗·阿拉维纳设计的新型社会住房,都是参与式设计、对社会尽责的建筑设计的典范。这些获奖者利用他们的建筑技能来解决事关全球的重要问题,并通过建筑设计创造机会,提高生活质量和引领未来。

多西的建筑理念对建筑师的角色有一个深刻的整体理解,对于居住者和占用空间的人来说,这是一座完全为印度风格的建筑,同时将当地的乡土传统和现代主义的意蕴融合在了一起。他的建筑设计对环境和地方风俗都进行了仔细的考虑,运用传统、地域性和人性化的现代主义原则进行探索性设计,并与当地的气候和环境相适应。

1927年,多西出生于印度普纳一个从事家具业的家庭。1951年,他前往巴黎为20世纪最伟大的现代主义艺术家勒·柯布西耶和路易·康工作,后来才回到孟买学习建筑设计的。1954年,他回到印度,负责柯布西耶在昌迪加尔市和艾哈迈达巴德市的项目,包括工厂主协会大楼和Shodhan别墅。在艾哈迈达巴德,他留了下来,并建立了自己的事务所,起名叫Vastu-Shilpa,是以印度的一家建筑知识工作室的名字命名的。1962年,多西与路易·康合作创办印度管理学院,并一直持续到20

construction. He endeavored to counteract this by using materials from demolished buildings as a way of crafting new architecture. Here is an architecture which is recomposed of memory, highlighting its craftsmanship, sense of time, and memories of people who inhabited such place.[1] Shigeru Ban's temporary refugee shelters for disaster and conflict zones and Alejandro Aravena's new typology of social housing, are examples of a participatory and socially responsible architecture. These laureates have utilized their architectural skills to tackle important issues that matter globally, and created opportunities through architecture that improve quality of life and shape the future.
Doshi's architectural philosophy lies with a deep holistic understanding of the role of the architect, for people who inhabit and occupy space, an architecture which is all Indian, whilst hybridizing local vernacular traditions with modernist overtones. His architecture is a careful consideration of context and local practices, exploratory design using traditional, regional and humanized modernist principals and formed to fit with the local climate and environment.
Doshi was born in Pune, India, in 1927 into a family that had been associated with the furniture industry. He studied architecture in Mumbai before making a one way trip to Paris in 1951 to work for the greatest modernists of the 20th century, Le Corbusier and Louis Kahn. He returned to India in 1954 to oversee Corbusier's project for the city of Chandigarh and Ahmedabad including the Mill Owners' Association Building and Villa Shodhan. At Ahmedabad, he stayed and established his own practice, Vastu-Shilpa, which bears its name from an Indian practice of architectural knowledge. In 1962, Doshi partnered with Louis Kahn on Indian Institute of Management, and continued to collaborate with him into the 1970s. In the spirit of Le Corbusier, and Louis Kahn, Doshi gained independence in creativity, freedom

"CEPT校园立即有了小小的校园和一座大房子,让每个人都意识到自己在其中的角色。学习是同时发生的,我们对环境的响应塑造了我们的生活。"最初,校园内只容纳了建筑学院(1966年),后来扩大到包括规划学院(1970年)、视觉艺术中心(1978年)、建筑科学技术学院(1982年)、室内设计学院(1982年)、卡诺利亚艺术中心(1984年)和展览馆(2012年)。CEPT大楼可以响应每个学科的不同需求,这个不断发展的校园为持续扩建提供了空间。教职员工和学生的往来则保持着身体上和环境上的相互联系。

"CEPT campus has become at once a small campus and a big house, making everyone aware about his or her role in it. Learning occurs simultaneously and our responses shape our lives." Initially housing just the School of Architecture (1966), the campus has expanded to include the School of Planning (1970), the Visual Arts Centre (1978), the School of Building Science and Technology (1982), the School of Interior Design (1982), Kanoria Centre for Arts (1984), and an exhibition gallery (2012).
CEPT buildings respond to the distinct needs of each discipline, and the evolving campus allows space for continued expansion. Crisscross movements of faculty and students maintain physical and environmental interconnectedness.

建筑学院的行政楼
School of Architecture from the Administrative block

环境规划技术中心,艾哈迈达巴德,印度,1966—2012年
Centre for Environmental Planning & Technology, Ahmedabad, India, 1966–2012

工作室下方的互动空间
Spaces for interaction below the studios

世纪70年代。本着勒·柯布西耶和路易·康的精神,多西获得了创作上的独立、表达上的自由,以及对物质和触觉的理解,并且通过气候环境、可持续性和以人类社会为中心的原则来探索设计概念。

根据多西的观点,印度认为建筑并不是一种独立于生命的形式,一切都与我们的身体、情感、生理机能和生活态度有关,对印度人来说,万事万物皆相关。²

在印度的艾哈迈达巴德、班加罗尔、海德拉巴和斋蒲尔等城市,有超过100栋值得一提的建筑,多西最著名的作品之一是阿兰若低成本住宅(1989年),它围绕着一群回旋盘绕的房屋进行设计,包括公共庭院、不同层次的蜿蜒环路和通道。这些住宅的设计考虑到了扩建和改建的适应性,它提供了一系列的可能性,从单人间的庇护所到相对宽敞的大房子。这是一个富有创新精神的场所,容纳了各种不同的收入群体,在一个空间复杂的项目中扩展了公共领域和各种公共层次的概念。这是根植于当地情感和社会观念的建筑设计项目。

作为一名实践者、教育家,多西于1962年在艾哈迈达巴德的一处荒地上创建了CEPT大学,这是一所建筑大学。他采用了三大原则: 1.土地的改造; 2.教育不该有界限,不该有壁垒,更不该有任何限制; 3.利用勒·柯布西耶的原则来利用气候和微风。它发展成为一个建筑体系,利用自然的微风和树木遮阳和降温,在空间中形成土地、气候和运动。一系列内外相互连接的空间和楼层创造了一层又一层多用途、正

of expression, and understanding materiality and tactility, exploring concepts through climatic context, sustainability and human society centric.
According to Doshi, India does not think that building is a separate form from life, everything is connected to our body, emotions and physiology and attitude to life, and that for Indians everything is related.²
With over 100 buildings credited across Indian cities such as Ahmedabad, Bangalore, Hyderabad and Jaipur, one of Doshi's most notable works is the Aranya Low Cost Housing (1989), which is designed around a cluster of convoluted houses, comprised of communal courtyards, diverse layered and meandering interlocking pathways and passages. The dwellings are designed with extension and adoptability in mind. It presents a range of possibilities, from one-room shelters to comparatively spacious houses. This is an innovative site and accommodates a variety of mixed income groups, expanding the public realm and the concept of various communal layers in a spatially complex project. Here is an architecture which is rooted in local sensibilities and social outlook.
As a practitioner and an educator, Doshi founded the CEPT University in 1962, an architecture University which he designed on a wasteland in Ahmedabad. There were three main principles that he adopted; 1. Transformation of the land, 2. Education should have no boundaries, no walls, no structures which can limit the space and 3. Utilizing the principles of Le Corbusier to use climate and breeze. This developed into an architecture which is designed formulating the land, climate and movement in space utilising natural breeze and trees for shading and cooling. A series of inside-to-outside interconnecting spaces and levels created layers of multi-use, formal, informal and intimate spaces; a porous symphony of spaces in constant flux.

卡马拉住宅，艾哈迈达巴德，印度，1963年
Kamala House, Ahmedabad, India, 1963

Sangath Architect的工作室，艾哈迈达巴德，印度，1980年
Sangath Architect's Studio, Ahmedabad, India, 1980

印度学研究所，艾哈迈达巴德，印度，1962年
Institute of Indology, Ahmedabad, India, 1962

Vidhyadhar Nagar总体规划，斋蒲尔，印度，1984年
Vidhyadhar Nagar Masterplan, Jaipur, India, 1984

对昌迪加尔和老斋浦尔进行分析，构建一个考虑交通、人口、就业模式、基础设施和环境利用的社区。
"我们能尽量减少机动车出行，尽量多走路和骑车吗？我们可否改变传统的土地利用方式，并提供混合土地用途？我们能否以不同的尺度创造机会让人人都能在一起工作？我们能找到最理想的运输网络吗？"这些是建筑师在设想住宅综合体的总体规划和城市设计时探索的一些问题。

Chandigarh and Old Jaipur were analyzed to construct a community that considered transportation, demographics, employment patterns, infrastructure and environmental use.
"Can we minimize motorized travel and maximize walking and cycling? Can we change the conventional land use patterns and provide for mixed land uses? Can we create opportunities at various scales where everybody can work together? Can we find an optimal transportation network?" These were some of the questions explored by the architect as he envisioned the masterplan and urban design for the housing complex.

式、非正式和亲密的空间，创作了一首不断变化的通透空间交响曲。

建筑中的标志性楼梯由两座相连的桥设计而成，他将这个楼梯描述为做游戏，做一个建筑游戏。3

多西在过去的十年中最引人注目的标志性项目之一，就是班加罗尔的印度管理学院（1977—1992年），它受到了印度各地如迷宫一般的传统城市和寺庙的影响。它融合了相互连接的建筑、庭院和画廊、半开放的大型走廊和绿色植物，当人们在建筑物和空间中移动的时候，居民也可以同时在室内和室外活动。它包含了一系列宝塔柱、开放和封闭的中间多孔连接，以及许多带有庭院空间的教室，不断地演绎着光与影的故事。

在逾70年丰富多彩的职业生涯中，巴克里希纳·多西的作品反映了他在其所处大环境下对建筑的真实理解和建筑的艺术性，包括各个社会经济阶层、各种类型的建筑。通过抑扬顿挫、深思熟虑的强化，他的建筑呈现出一种高贵的诗意，表达了一种对历史、集体文化价值和人文精神的更深刻的理解，对印度乃至全世界的建筑教育都产生了深远的影响。

"我的作品就是我的生命、人生观和梦想的延续，试图创造建筑精神的宝库。我要把这一殊荣归功于我的导师勒·柯布西耶。正是他的教导使我开始质疑身份，并迫使我发现适应本地情况的全新当代表达方式，从而实现可持续发展的整体栖息地。"多西说。

Its iconic staircase was designed with two bridges which interconnect, which he described as playing a game, a game of architecture.[3]
One of Doshi's most compelling and iconic projects over the past decade is the Bangalore's Indian Institute of Management (1977-1992) which was influenced by traditional maze-like Indian cities and temples which are found across India. It is a collection of interlocking buildings, courts and galleries, semi-open large corridors and greenery which allow inhabitants to be concurrently indoors and outdoors, as people move throughout the building and spaces. It contains a series of Pagoda columns, open and closed in-between porous connections and many classrooms with courtyard spaces that provide a constant play between light and shadow.
Balkrishna Doshi's body of work reflects a true understanding and artistry of architecture in its context, from every socio-economic class across a broad spectrum of genres over a rich 70-year career. His architecture displays a distinguished poetry which expresses a deeper understanding of history, cultural values of collectives and a human spirit through dactylic and thoughtful reinforcement which has had a lasting influence on Architecture education in India and across the world.
"My works are an extension of my life, philosophy and dreams trying to create treasury of the architectural spirit. I owe this prestigious prize to my guru, Le Corbusier. His teachings led me to question identity and compelled me to discover new regionally adapted contemporary expression for a sustainable holistic habitat." said Doshi. Gihan Karunaratne,

Images and photographs courtesy of VSF

1. Architect Wang Shu, 2012 Pritzker Winner, Washington University in St. Louis
2. Premjit Ramachandran, dir. *Doshi: Architecture Without Adjectives*, Film, Hundredhands, 2007. Transcript, DOSHI Chapter 2 - "I'm not an architect for me it's a search"
3. Balkrishna V. Doshi: 27th Annual Architecture Lecture, Royal Academy of Arts, 2017.

伊丽莎白码头大桥
Elizabeth Quay Bridge _ Arup Associates

伊丽莎白码头大桥是一座优雅的步行和自行车桥,横跨澳大利亚珀斯中部的天鹅河入口。这座桥实际上是通往中央商务区的门户,其曲线外形在视觉上将河流与城市联系起来。

该概念设计基于设计团队的愿望,他们希望实现一种能考虑到总体规划的桥梁形式,在利用其位置优势的同时实现美学和功能方面的要求。共享建模的使用对于项目的成功交付是至关重要的,建筑师和工程师都使用Rhino + Grasshopper脚本来控制几何定义——每天共享的脚本确保了结构分析与建筑的几何形式得以保持同步。

设计团队最后交付的桥面边缘深度只有250mm。这种优雅的设计超出了客户的审美预期,体现了团队对项目所采用的创新建筑设计和工程方法。这座大桥从各个角度看去都很具有吸引力。110m长的S形桥面包含了清除航道所需的必要长度,同时也为行人和骑自行车的人提供了充满活力、不断变化的视角。这种建筑形式在天鹅河上创造了自然的有利位置,回过头来,又将伊丽莎白码头的中央和珀斯的中央商务区当成了背景。无论白天还是夜晚,这座大桥都是当之无愧的焦点,每当夜幕降临,它都会与集成功能照明装置在码头附近共同营造一个精致的环境、一个令人放松的氛围。

伊丽莎白码头大桥拥有引人注目的建筑特征,其简单而纯粹的结

构设计形式满足了客户的所有需求，辅以合理的模块化建筑设计，以及饰面、栏杆和木板桥面。该设计方案克服了每一项技术和施工方面的挑战，同时还能使桥梁的美观要求保持在设计的最前沿。

The Elizabeth Quay Bridge is an elegant pedestrian and cycling bridge that spans the inlet of the Swan River in central Perth. The bridge is effectively a gateway to the central business district, its curvaceous form visually linking the river with the city.

The concept design was driven by the design team's desire to achieve a form for the bridge that respected the masterplan, and used its location to advantage, while achieving the aesthetics and functional briefs. The use of shared modelling was pivotal to the successful delivery of the project, with both architects and engineers using Rhino + Grasshopper scripts to control the geometry definition – scripts that were shared on a daily basis to ensure that structural analysis kept pace with architectural geometry.

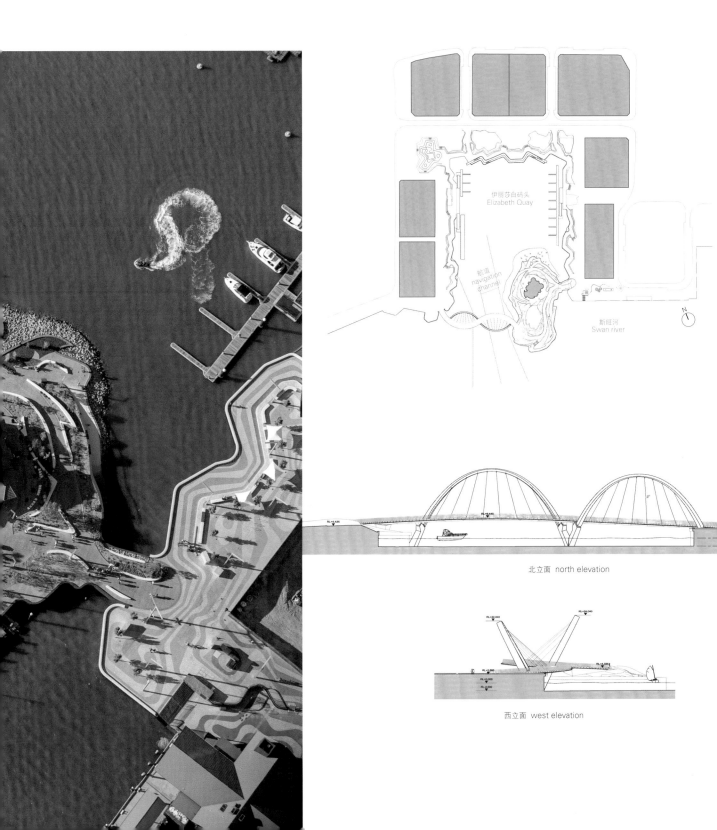

北立面 north elevation

西立面 west elevation

The design team delivered a bridge deck that was only 250mm deep at the edges. This elegance exceeded the client's aesthetic expectations and is a manifestation of the team's innovative architectural and engineering approach to the project. The bridge is appealing and attractive from every angle. The "S" form of the 110m long deck incorporates the necessary length required to clear the navigation channel while providing dynamic and changing viewpoints for pedestrians and cyclists using the bridge. This form creates natural vantage points out to the Swan River and back towards the heart of Elizabeth Quay with the Perth CBD as a backdrop. Day and night it acts as a focal point, with integral feature lighting creating a relaxing and sophisticated ambience on the quay when night falls.

A striking architectural feature, the design of the Elizabeth Quay Bridge is everything the client sought to achieve in terms of simple and pure structural forms, complemented by a rationalised modular architecture to the fascia finishes, balustrading and timber decking. The design solution overcame every technical and construction challenge while ensuring the bridge's visual appeal was at the forefront of the design.

项目名称：Elizabeth Quay Bridge / 地点：Perth, Australia / 建筑师：Arup Associates / 作者：Alistair Avern-Taplin, Nick Birmingham
项目团队：Arup–Andrew Allsop, Alistair Avern-Taplin, Nick Birmingham, Peter Burnton, Stewart Buxton, Efren Cerrero, Anthony Ferrau, Anastasia Fragoulis, Kathy Franklin, Ruby Heard, Milena Kovac, Mira Lee, Michael Lin, Angus Low, Lewis MacDonald, Georgie Prie, Anu Ramachandran, Clayton Riddle, Nic Scanlan, Ed Spraggon, Julia Summers, Kai Tan, Jess Watts / 总承包商：CPB (formerly Leighton) and Broad / 客户：Perth Metropolitan Redevelopment Authority (MRA) / 面积：450m² bridge deck / 高度：22m
设计时间：2012.3.7 / 竣工时间：2015.9.1 / 摄影师：courtesy of the architect - p.11; ©Dion Photography (courtesy of the architect) - p.12~13; ©Jacaranda Photography (courtesy of the architect) - p.8~9, p.10~11 left

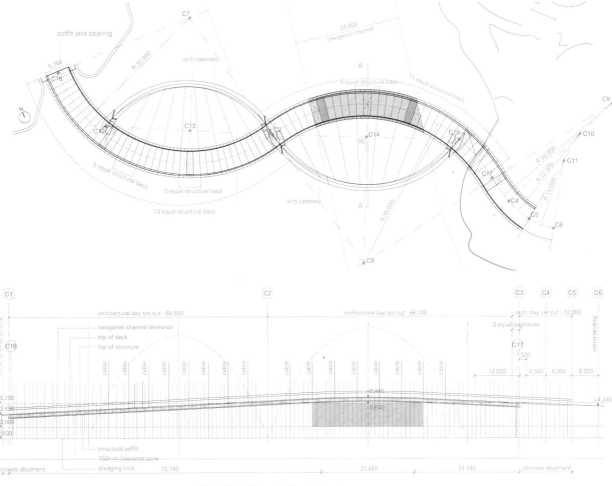

桥面中心线 bridge centerline deck level

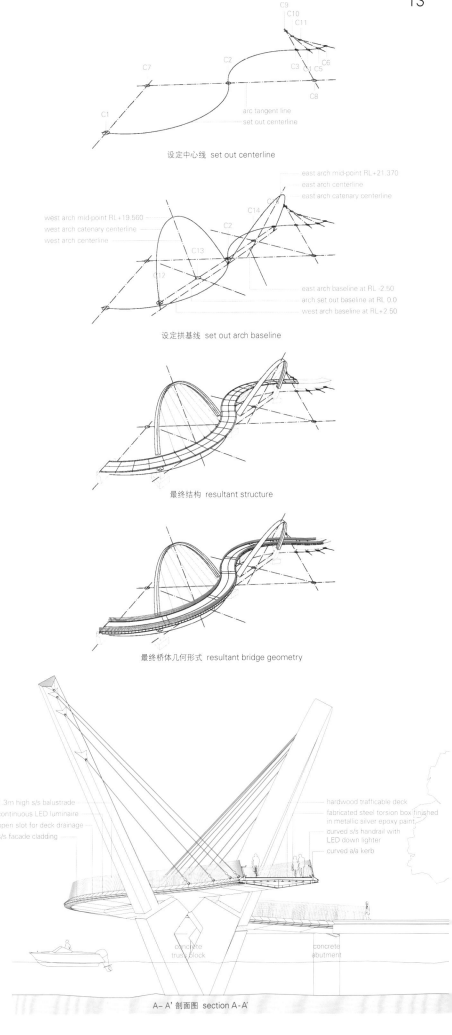

设定中心线 set out centerline

设定拱基线 set out arch baseline

最终结构 resultant structure

最终桥体几何形式 resultant bridge geometry

A-A' 剖面图 section A-A'

步行观景桥 Bridge for Walk and View

姆扎巴大桥

Mzamba Bridge _Carinthia University of Applied Sciences + buildCollective

姆扎巴大桥满足了当地居民安全渡过南非姆扎巴河的愿望。当时,这条河将居民与教育设施、医疗保健和普通食品供应等必要的当地基础设施隔开了。人们通常无法及时获得医疗服务,老年人也无法前往当地的零售市场。

从设计的角度来看,桥梁应该始终与周围环境保持很强的亲和力,但在这种情况下,人行桥的社会根源在于它的极端。需要考虑到的影响包括不便于通行(桥梁的所有构件都需要人工搬运到施工现场),以及与大多数缺乏专业技能的当地社区工人合作施工(提供工作积极性很高的劳动力)。

设计本身是从对可能入选的几个场所进行彻底调研开始的。位于场地一侧的一些合理的接触点(理想的地质岩层)和另一侧的小型非本地植被都是值得推荐的入选位置,并在当地的姆扎巴河支流中进行浇筑。为了避免使用重型机械,设计师将13m高的吊架安装在地面上,并按照五月柱(各类欧洲民族节日的组成部分)的方式搭建,没有安装脚手架。

在上部结构中,由使用者和侧向风引起的振动(悬索桥的关键之处)通常通过使用更多的材料,也就是增重,来解决,有助于减少不必要的振动。在设计过程中,设计师开发了预应力钢索系统,大大增强了桥梁的刚度,使桥梁显得更加稳重。设计师使用预应力锚索形成了一条拱形走道,为浮木和漂流到河岸的货物留下了足够的空间。对于外行人来说,设计的其他优势仍然是难以察觉的:整座桥是自行开发而成的,所有施工所必需的构件也都是建成桥梁的一部分。一方面,这使得现场的施工工作更加轻松,也保证了在维修过程中更容易更换个别的构件。

The Mzamba Bridge deals with the local population's desire for a safe crossing over the river Mzamba in South Africa. The river used to separate the inhabitants from necessary local infrastructures such as educational facilities, healthcare and general food supply. Often medical care couldn't be reached in time and the elderly had no means of accessing the local retail market.

From design point of view, a bridge should always have a strong affinity to its surrounding environment, but in this case the footbridge roots socially in its extreme. Effects such as the difficult accessibility (all parts of the bridge need to be carried to the construction site manually) and construction in cooperation with the mostly unskilled local community (providing highly motivated manpower) needed to be taken into account.

The design itself started with a thorough survey of potential locations. Some logical contact points (an ideal geological rock formation) on one side and small non-native vegetation on the other site have recommended the chosen location and are casted with the local river from the Mzamba river. In order to avoid heavy machinery, the 13m pylons have been assembled on the floor and put up in the way Maypoles are erected as a part of various European folk festivals – without scaffold.

In the upper structure, oscillations (the most critical aspect for suspension bridges) caused from users and / or lateral winds are normally solved in form of more materials – more mass – that helps reducing unwanted oscillations. During the design, a system of prestressed steel cables is developed to increase the stiffness considerably, making the bridge as if it is heavier. This idea to use prestressed cables leads to an arched walkway leaving enough space for driftwood and jetsam. Other things remain imperceptible for the layperson: The whole bridge develops out of itself, and all components that are necessary for construction are also part of the finished bridge. This makes on the hand the construction easier and guarantees easier replacement of single pieces during maintenance.

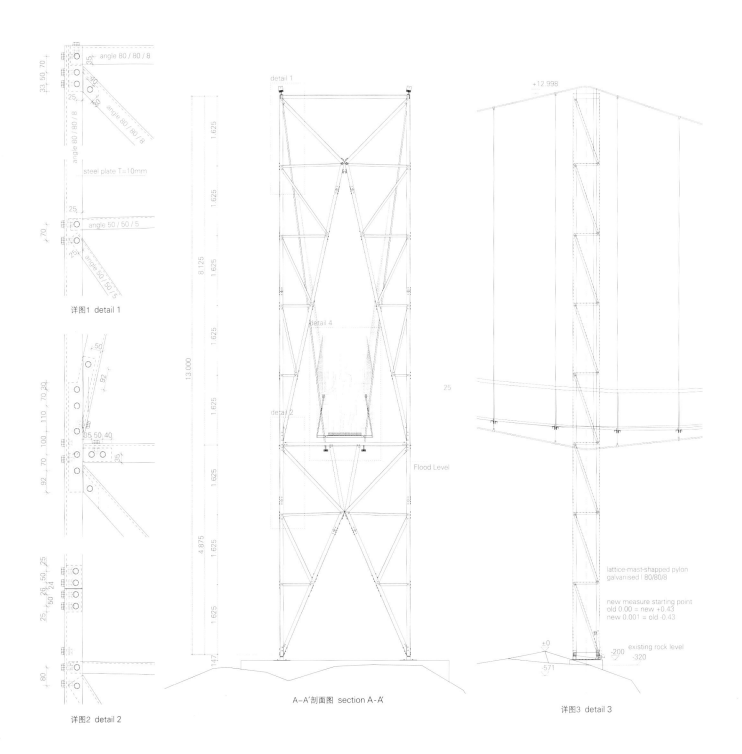

详图1 detail 1

详图2 detail 2

A-A'剖面图 section A-A'

详图3 detail 3

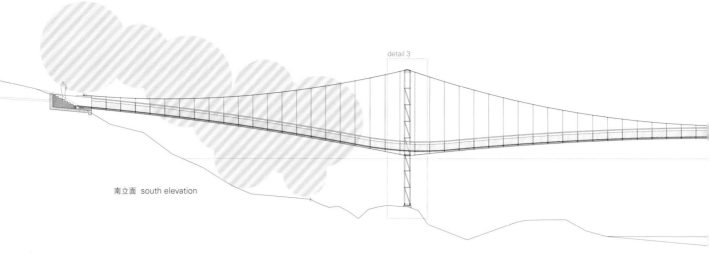

南立面 south elevation

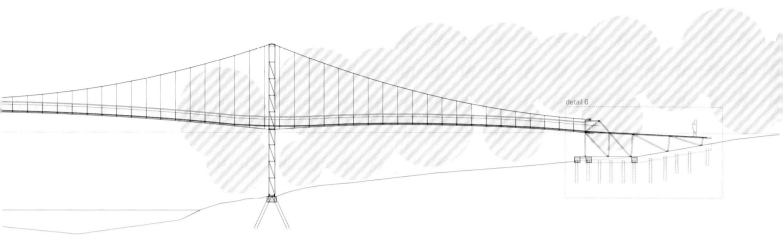

项目名称：Mzamba Bridge
地点：Mzamba River, Eastern Cape, South Africa
建筑师：Thomas Harlander, Florian Anzenberger
项目团队：Thomas Harlander, Florian Anzenberger, Marlene Wagner, Elias Rubin
项目管理：Marlene Wagner, Elias Rubin
合作方：Carinthia University of Applied Sciences + buildCollective
结构工程师：Lüchinger Meyer Ingenieure, Rudi Keudel Schaeffer, Bruce Plumbly
施工：Project team with people living around Mzamba river
客户：Mzamba steering committee founded by locals living around the river
用途：pedestrian bridge / 长度：132m
结构：suspension bridge
外部饰面：galvanized steel frames with wooden handrails and wooden walkway
材料：concrete for the foundations, upper structure with steel and handrail + walkway in wood
造价：200,000 EUR, including material donations and labour contribution
竣工时间：2015.10
摄影师：©bridgingMZAMBA (courtesy of the architect)

详图4
detail 4

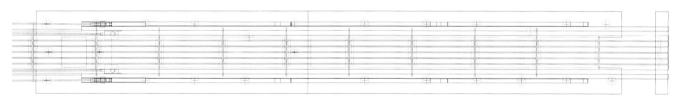

详图5
detail 5

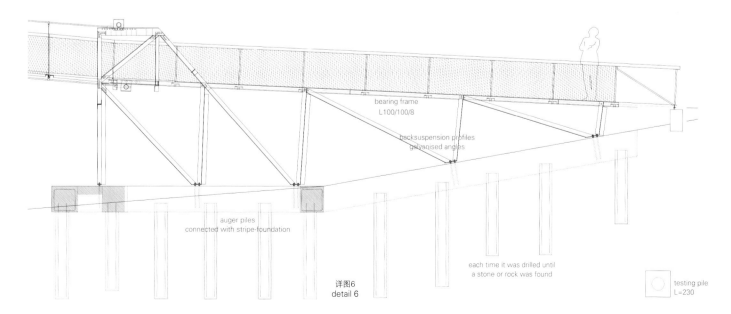

详图6
detail 6

步行观景桥 Bridge for Walk and View

斯蒂尔树顶人行道
Stihl Treetop Walkway _Glenn Howells Architects

Westonbirt植物园是世界上最好的树木收藏地之一。它拥有16000个独特的树种,包括一棵拥有2000年历史的菩提树,所有树木均精心布置在树林、游乐设施、林间空地和丘陵地上,形成了迷人的一级列管历史景观。

斯蒂尔树顶人行道提供了令人叹为观止的新观景视角,特别是古老的巨盘木林地和丘陵地,达到了给该地块自然遗产升值的效果。它几乎有300m长,是英国同类建筑中最长的,不但提高了植物园的形象,也带来了来自全英国各地的新游客。

这条步道从丘陵地的南端开始,穿越巨盘木林地,最后止于一块名为"协和林间空地"的空地。这条路线充分利用了当地的自然地形,在山谷间架起了桥梁,使人们在没有楼梯或电梯的情况下,也很容易到达地面的人行道。

这将使游客能够沿着桥梁建筑毫无阻碍地行走,走着走着,山谷在他们的下方消失了,人行道轻松地攀升到森林地面以上13.5m处。这种设计方法是经过深思熟虑的,它将向更多的游客,特别是那些步履不便的游客,开放巨盘木林地,使他们都能够欣赏森林的美景。

这条路线蜿蜒曲折,拖曳至树冠上方,再穿过间距为10.5m、优雅的剪切状木支腿支撑的树冠,木支腿使建筑悬在空中。在路线的四个点上,人行道也从宽度为1.9m慢慢扩大到3.7m,为人们提供了在周围林地停歇、反思和演出的空间。

在设计中,设计师参考了树木的遗传密码,人行道的灵感来自于树木基本结构背后的复杂性和优雅性。然后,用参数化原理表示,参数化原理可以利用算法和参数将单线条的图形转化为错综复杂的设计模型。

利用这种参数化方法,设计师可以进行更大规模的试验,最终确保设计更加优雅、天衣无缝。经过大量的测试,支腿的间距设置为

10.5m，保证了视觉上的连续流动，跨越道路上方，并且最大限度地减少对现有树木的影响。

　　该项目始终追求卓越，在材料方面，由于采用了敏锐的设计方法和材料选择方式，建筑结构更加迎合环境。人行道是用优雅老化的落叶松建造的，落叶松风化成银灰色之后，便与钢栏杆相得益彰。

Westonbirt Arboretum is home to one of the finest tree collections in the world. It holds 16,000 unique specimens including a 2,000-year-old lime tree, and all trees are carefully laid out in woods, rides, glades and downs that form a beautiful Grade I listed historic landscape.

The Stihl Treetop Walkway provides stunning new views over this landscape, in particular the ancient woodlands of Silk Wood and across The Downs, enhancing the appreciation of the estate's natural heritage. At almost 300m in length, it is also the longest structure of its kind in the United Kingdom, raising the arboretum's profile and bringing new visitors from all over the country.

The walkway begins at the southern end of The Downs, sweeping over Silk Wood and finishing in a clearing called Concorde Glade. This route makes the best of the land's natural topography by bridging across a valley, allowing for ease of accessing to the walkway at ground level without any stairs or lifts. This will enable visitors to walk without hindrance along the structure while the valley falls away beneath and the walkway rises effortlessly over 13.5m above the forest site. This approach was deliberate and it would open up Silk Wood to more visitors, particularly those with mobility issues, enabling them to enjoy the forest.

The route takes a sinuous form, snaking above and through the tree canopy supported by elegant scissoring timber legs spaced at 10.5m intervals that hold the structure in the air. At four points along the route, the walkway also gently widens from 1.9m to 3.7m, providing spaces for pause, reflection, and interpretation of the surrounding woodland. Referencing the genetic code of trees in its design, the walkway is inspired by the complexity and elegance behind the fundamental structures of trees. This is then expressed with parametric principles, which can transform a single line into a detailed design model using algorithms and parameters.

This parametric approach enabled greater experimentation, ultimately ensuring a more elegant and seamless design. After considerable testing the leg spacing was set at 10.5m allowing for a visually continuous flow, spanning roads and minimising the impact on existing trees.

Striving for excellence throughout, in terms of materiality the structure ingratiates itself with its environment, thanks to a sensitive approach to design and material selection. The walkway is constructed of larch which ages gracefully, the timber weathering to a silver-grey to complement the steel balustrade.

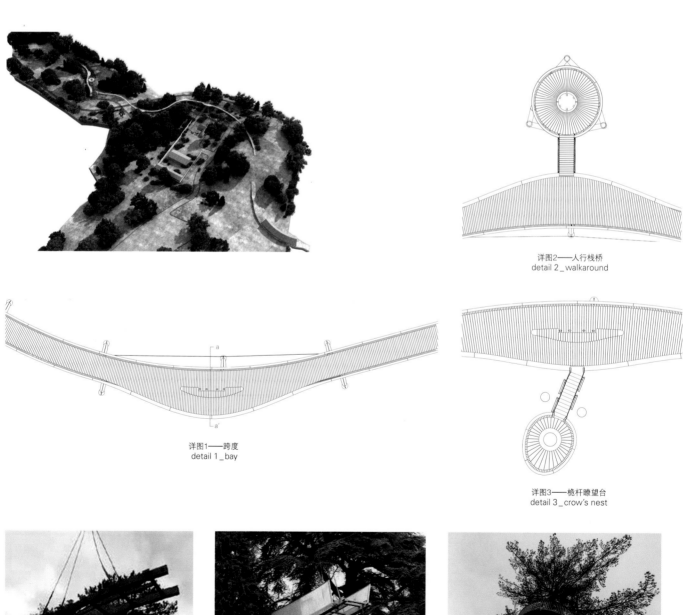

详图1——跨度
detail 1 _ bay

详图2——人行栈桥
detail 2 _ walkaround

详图3——桅杆瞭望台
detail 3 _ crow's nest

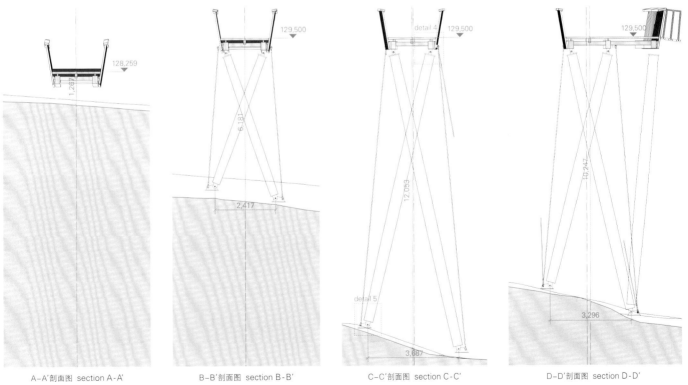

A-A'剖面图 section A-A'   B-B'剖面图 section B-B'   C-C'剖面图 section C-C'   D-D'剖面图 section D-D'

项目名称：Stihl Treetop Walkway / 地点：Westonbirt, The National Arboretum, Tetbury, Gloucestershire GL8 8QS, UK / 建筑师：Glenn Howells Architects
结构工程师：Buro Happold / 工料测量师：PMP Consultants / CDM-C顾问：PMP Consultants / 解说设计：Outside Studios / 总承包商：Speller Metcalfe
钢结构分包商&地上工程：SH Structures / 木柱工匠：Ventis & Brasker Masten / 木桥面板&栏杆：CTS Bridges, Russwood
桥面&栏杆处理：Koppers Micropro by Norclad / 绳索供应商：Bristol Rope & Twine / 钢丝网：Carl Stahl / 长椅制造商：Heseltine Design
造价：£1.7M / 尺寸：284m long, 1.9 to 3.7m wide, up to 13.5m high / 采购：Traditional contract / 客户：Stihl Treetop Walkway
竞赛获胜和委托时间：2010 / 设计时间：2014.8 / 施工时间：2015.4 / 竣工时间：2016.4 / 摄影师：©Rob Parrish (courtesy of the architect)

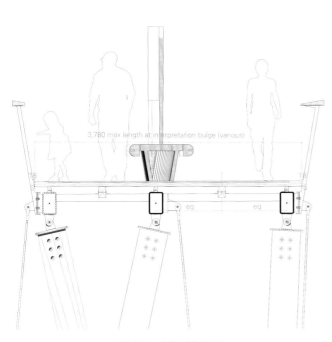

a-a'剖面图——表演区的短剖面图
section a-a'_short section through interpretation bay

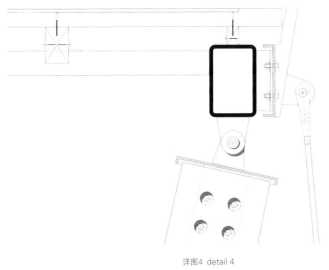

详图4 detail 4

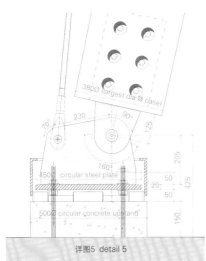

详图5 detail 5

维特雷P.E.M交通枢纽项目

P.E.M Vitré _TETRARC

作为一个大型城市开发项目,位于法国维特雷的P.E.M多模式交通枢纽就好像是城市中的嫁接项目,旨在简化使用、合并功能,并衔接各个区域。为了达到这个目的,它吸收了周围环境的组成部分,确保能利用到附近所有重要的功能。

位于维特雷车站的行人天桥在城市里由北向南横跨铁路,可到达所有的月台。

第二座人行天桥的下方是一个有620个车位的地下停车场,这座天桥将第一座人行天桥和维克托瓦尔广场连接起来,从而创建了从这个广场到北站停车场的步行路径。由于停车场容量极大,所以它必须拥有强大的建筑结构,大约130m长,近10m高。

这座桥将整个停车场延伸到场地的东端,将其分成一系列连续的凹槽,以呼应场地的主要形态节奏。它的体量与所有屋顶的绿色开发项目相结合,因为它可以融入目前南侧悬崖的绿色自然环境中。

从上层停车场开始,人行通道不仅被划分为易于使用电梯、楼梯的功能序列,也被认为是一段长达180m、放飞心灵的旅程,在路上设有许多有吸引力的观景点,可以欣赏维特雷古堡以及整座古城,反过来再逐渐近距离欣赏南山坡的景观和地标建筑。

维克托瓦尔广场的上半部分沿着停车场的南立面与皮埃尔·勒迈特大街相连,背靠现有的山坡。

一个公共空间服务于地下停车场,有16个车位的地上停车场,通过人行天桥服务于圣玛丽学校和车站的落客空间,以及一套道路系统,环绕着指定用于未来建设办公楼的场所。

As a large urban development project, P.E.M, the multi-modal transportation hub in Vitré, appears to be an urban graft intended to simplify uses, combine functions, and join

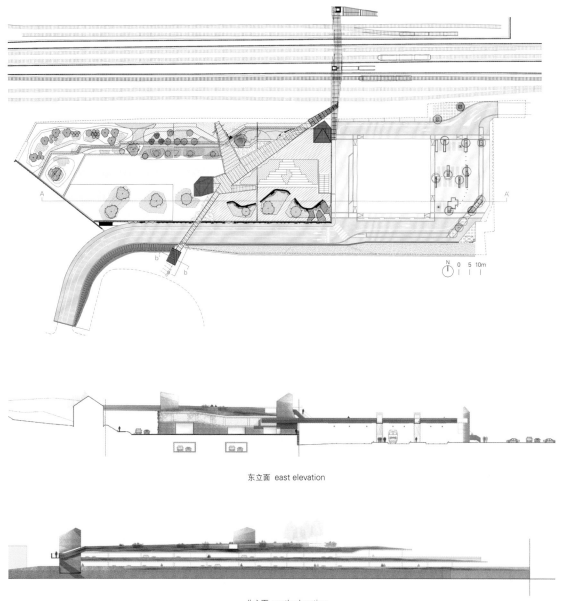

东立面 east elevation

北立面 north elevation

territories. To this end, it assimilates the components of its surroundings by ensuring that all vital functions are assigned to it.

A pedestrian footbridge in Vitré Station, spanning the railway from the north to the south of town and serving all platforms. A 620-space underground car park is covered by a second pedestrian footbridge connecting the first footbridge to the Place de la Victoire, thereby creating a pedestrian journey from this square to the Northen Station car park. Due to its large capacity, this car park must be a powerful object which is approximately 130 meters long and reaches almost 10 meters high.

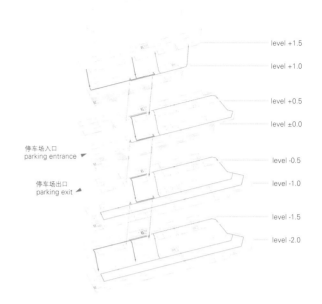

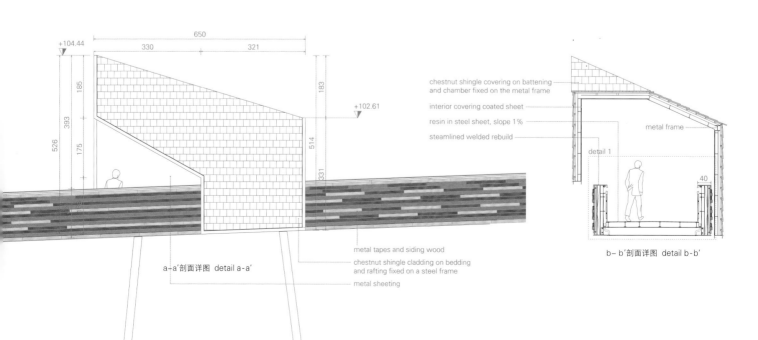

a-a'剖面详图 detail a-a'

metal tapes and siding wood
chestnut shingle cladding on bedding and rafting fixed on a steel frame
metal sheeting

chestnut shingle covering on battening and chamber fixed on the metal frame
interior covering coated sheet
resin in steel sheet, slope 1%
steamlined welded rebuild
metal frame

b-b'剖面详图 detail b-b'

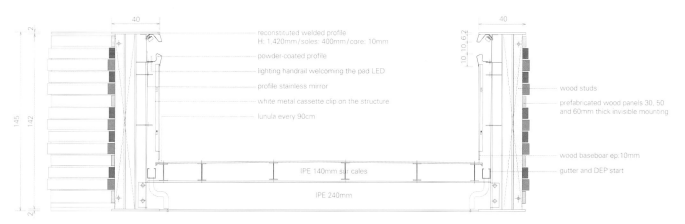

详图1——桥梁典型剖面 detail 1_bridge typical section

The bridge stretches the overall body of the car park to the site's eastern tip, breaking it down into a set of successive recesses to answer the dominant morphological rhythms of the site. The mass of this equipment is coupled with the green development of all its roofs, as it can be assimilated into the current green nature of the south cliff.

From the upper parking area, the pedestrian path is not only divided into functional sequences that ease the use of lifts, stairs, and stairways, but also considered as a psychologically fragmented journey to be crossed along the distance (180m), with many attractive points of view on the Castle of Vitré and the whole ancient city, inversely by gradually zooming in on the landscape and built landmarks of the south hillside.

The upper part of the Place de la Victoire is connected to rue Pierre Lemaître along the south facade of the car park, against the existing hillside.

A public space serves the underground car park, an above-ground car park with 16 spaces, drop-off spaces serving the St. Mary's School and the station via the footbridge, and a road system curving around the place earmarked for the future construction of an office building.

A-A'剖面图 section A-A'

项目名称：P.E.M Vitré
地点：City of Vitré (35), France
建筑师：TETRARC
总建筑师：Jean-Pierre Macé, Olivier Perocheau, Julie Goislot
结构工程师：BETREC, E2C + SCE
流体工程师：ICOFLUIDES
景观：Louise Folin
施工：Olivier Pérocheau, Jean-Pierre Macé et Julie Goislot
会计：Véronique Delezir
总建筑面积：17,664m²
SNCF大门长度：60m
人行桥总长度：160m
造价：16 M€ HT
桥梁竣工时间：2015
停车场竣工时间：2016.6
摄影师：
©Stéphane Chalmeau (courtesy of the architect) -p.30, p.31, p.33; ©Thomas Vittu (courtesy of the architect) -p.26~27, p.28~29, p.32

步行观景桥 Bridge for Walk and View

萨默斯镇大桥
Somers Town Bridge _Moxon Architects + Arup

　　Moxon Architects建筑师事务所与奥雅纳工程顾问公司合作,为国王十字合作集团(KCCLP)设计建造了一条崭新的人行天桥。这座桥横跨摄政运河,连接坎利街和新建的加什古花园,为西行通往萨默斯镇的行人和骑自行车的人提供了一条重要的通道。

　　整个跨度为38m的桥梁只有15mm厚,每一端的最小厚度为400mm。为了与附近维多利亚时代的伟大设计保持一致,这座桥坚固耐实,采用了精细的工艺和精挑细选的材料,力图营造一种极简主义的外观。这座桥的形式与其弯矩完全吻合,能直观地让人看到钢材都被用在了横梁的什么位置。从物质的意义上来说,这是"精益"设计。一条宽阔的坡道把人们带到了桥上,人们再从水面上方走过。

　　优雅的护墙从不锈钢过渡到刨花硬木,形状从梁的曲线轮廓到扶手完美的笔直线条;栏杆的材料得到了精心考虑,因为它们是桥梁和使用者之间最重要的实体连接。通过确定桥面板上方的结构深度,设计师让人们站在桥上就能清楚地欣赏到运河上的圣潘克拉斯水闸以南的美景:空气通风实现了最大化,从而确保人们能够看到运河上所有等待使用水闸的船只,而不会被桥梁遮住,并且保证能尽量清晰地看到圣潘克拉斯水闸和小屋的景致。

　　这座桥衔接了两个截然不同的场所:从东部煤渣场的复杂都市风,到西部城市边界附近的坎利街自然公园的天然去雕饰。它坐落于摄政王运河保护区内,周围都是列管建筑,包括圣潘克拉斯水闸。

Moxon Architects建筑师事务所的常务董事本·阿迪补充说:"这座桥的设计目标是在形式和材料使用方面实现'极简品格'。这是一次经过深思熟虑而又美观的设计,但更重要的是,我们开发的这种设计,跟其建于维多利亚时期的左邻右舍相比,同样强大、有趣而又实用。"奥雅纳的项目总监伊恩·威尔逊也补充说:"行人和骑自行车的人现在可以通过这种全新的连接方式轻松地往返于国王十字车站和卡姆登。"这座桥不但拥有复杂的几何结构和稳健的工程设计,修长而精致,并且呼应了基地的工业历史,因此它是国王十字路口开发项目最适宜的补充。

Moxon Architects, working with engineers Arup, have unveiled the new pedestrian bridge for the King's Cross Central Limited Partnership (KCCLP). The bridge spans the Regents Canal between Camley Street and the new Gasholder Gardens, forming an important connection for pedestrians and cyclists to the west, into Somers Town.

The entire 38m span of the bridge is only 15mm thick and achieves a minimum depth of 400mm at each end. In keep-

项目名称：Somers Town Bridge
地点：King's Cross, London, UK
建筑师：Moxon Architects
项目团队：Ben Addy, Tom Budd, Adam Holicska
结构工程师：Arup - Ian Wilson, Martin Hooton, Joan Valls Mestre
照明工程师：Hoare Lea - Gonzalo Pastor Peñalba, Phil Summers
景观设计师：Townshend Landscape Architects - Andrea Dates; Applied landscape design - Kym Jones, Claire Walsh
设计经理：Carillion - David Crowley
总承包商：Carillion
客户：King's Cross Central Limited Partnership (KCCLP)
用途：Infrastructure / Public realm
桥梁：38m span of the bridge is only 15mm thick and achieves a minimum depth of 400mm at each end
总建筑面积：364m²
结构：through plate girder
材料：stainless steel
设计时间：2014—2016
施工时间：2016
竣工时间：2017
摄影师：©John Sturrock (courtesy of the architect)

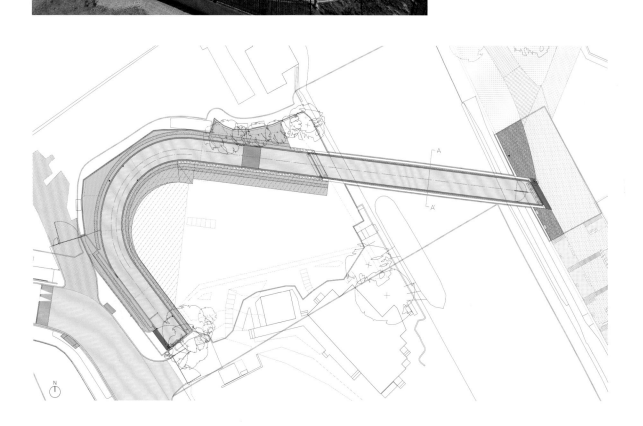

ing with the great Victorian design in the surrounding vicinity, the bridge is robust, with detailed craftsmanship and a careful selection of materials for a minimalist finish. The form of the bridge precisely matches its bending moment, making it a direct demonstration that where the steel is required in the beam. In a material sense it is a "lean" design. A sweeping ramp leads people up to the bridge and over the water. An elegant parapet transitions from stainless steel to planed hardwood, changing in shape from the curved profile of the beam to the perfectly straight line of the handrail; the railing materials are especially considered as they form the key physical connection between the bridge and the user. By locating the structural depth above the deck level, the design maintains a clear view of the canal south from St. Pancras Lock: the air draught is maximised, ensuring that any canal boat waiting to use the lock can be surveyed without the bridge obscuring them, as well as keeping views to the St. Pancras Lock and Cottage as open as possible.

The bridge connects two very different sites: from the sophisticated urbanism of the Coal Drops Yard to the east, to the natural and wild Camley Street Natural Park at the western abutment. It is located within the Regent's Canal Conservation Area, and surrounded by several listed structures, including the St. Pancras Water Tower.

Moxon Architects' managing director Ben Addy, added: "The ambition for this bridge has been to achieve 'extreme simplicity' in terms of form and material use. It is a considered and beautiful addition to the location, but even more than this we have developed a design that is as emphatically and enjoyably practical as its Victorian neighbours." Ian Wilson, the project director at Arup added: "Pedestrians and cyclists can now enjoy travelling to and from King's Cross and Camden with ease by using this new connection. It is a fitting addition to the King's Cross development as the complex geometry and robust engineering have resulted in a slender and refined bridge, which also echoes the site's industrial past."

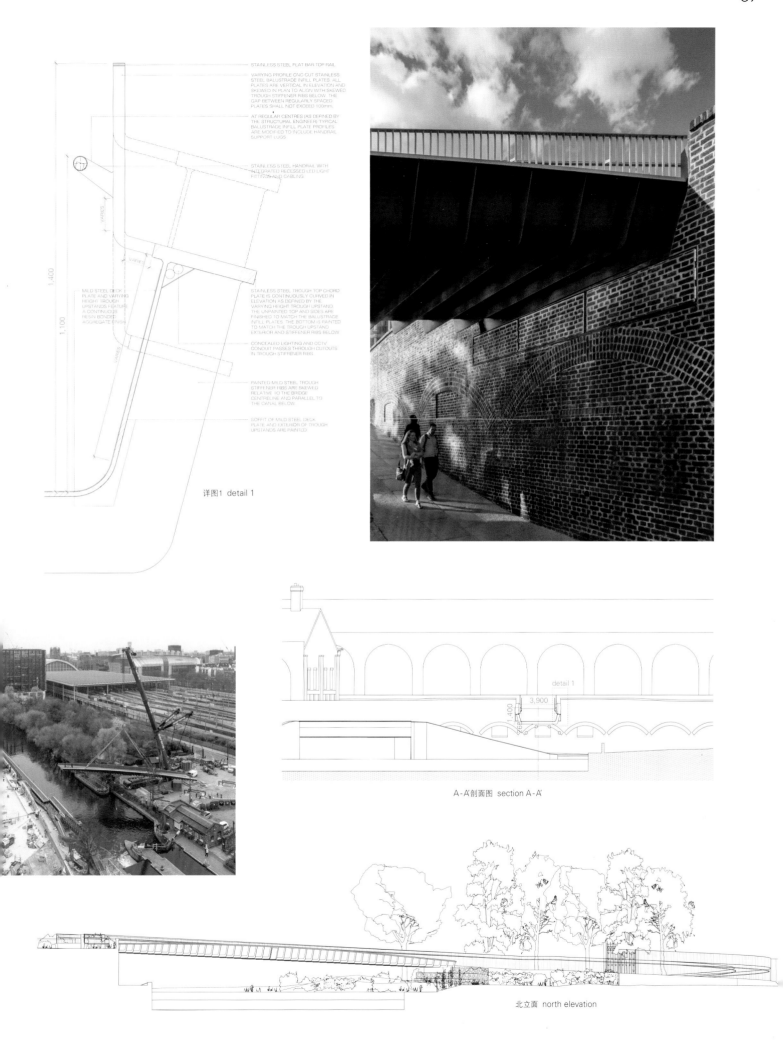

详图1 detail 1

A-A'剖面图 section A-A'

北立面 north elevation

作为抵抗方式的文化

在一个所有人类价值都被疯狂商品化的时代,提倡文化听起来就像是徒劳无益地反对资本的力量,这些力量继续不受干扰地侵蚀任何没有直接货币价值的东西。同样的道理也适用于另一个词:城市。当一切因后果都可以归于城市的细枝末节之时,谈论城市就越来越远离某种将社会维系在一起的社会纽带的想法了。然而,如果我们看看"文化"和"城市"这两个词所遵循的词源学轨迹,就可以开始设想,这两个词是如何包含着与另两个经常被误认为同义词的词(文明和城市化)对立的元素的。事实上,20世纪的建筑和城市主义的历史正是由这一组术语及其各自的倡导者之间的持续斗争所塑造的。从这个意义上说,文化可以被视为一种抵抗策略。

"文化"一词被视为对以市场为导向的城市化的不断波动和不确定性的一种反应,它给与之有关的建筑增加了一种通用的特性描述,即有可能仅仅把它们变成多功能的容器。

In an age of rampant commodification of all human values, advocating culture might sound like just one more word used in vain against the forces of capital that continue undisturbed in their action of erosion of anything that does not have an immediate monetary value. The same could be said for another term, city. When anything can be ascribed to the capillary condition of the urban, speaking of a city is increasingly far from the idea of some sort of social bond holding together a society. Yet, if we look at the etymological trajectory followed by the terms culture and city, we can start envisaging how both contain an element of opposition against two other words that are often mistaken as their synonyms: civilization and urbanization. In fact, the history of the 20th-century architecture and urbanism has been shaped by the constant battle between those couples of terms and their respective advocates. In this sense, culture can be considered a strategy of resistance.
Seen as a response to the constant fluctuations and uncertainty of market-driven urbanization, the adjective "cultural" adds a generic characterization to the buildings to which it is attached, with the risk of turning them into mere containers for multiple functions.

塞尔维亚卡伊莎文化中心_CaixaForum Sevilla/Guillermo Vázquez Consuegra
贝尔工作室——加拿大国家音乐中心所在地
Studio Bell, Home of the National Music Center/Allied Works Architecture
鹤岗文化大厅_Tsuruoka Cultural Hall/SANAA
麦德林现代美术馆扩建_Medellin Modern Art Museum Expansion/51-1 + Ctrl G
塞纳河音乐厅_La Seine Music Hall/Shigeru Ban Architects Europe
作为抵抗方式的文化_Culture for Resistance/Francesco Zuddas

Culture for Resistance

Francesco Zuddas

　　"然而，如果说今天的文化不再被赋予它在资本主义早期曾享有的某种程度上的相对自主权……这并不一定意味着它的消失或消亡。恰恰相反，我们必须继续确认这一点，自治的文化领域的解体可以被想象为一种爆炸：在整个社会领域内惊人的文化扩张，涉及我们的社会生活中的一切——从经济价值和国家权力，到实践，再到精神本身的结构——据说都可以成为某种原始的、尚未建立理论的意义上的'文化'。"[1]——弗雷德里克·詹姆逊

　　浏览本书中建筑师对项目的描述，我们会发现，他们在模糊边界方面有着共同的兴趣。塞纳河音乐厅（102页）是一个集音乐制作和表演于一体的空间，它坐落于法国塞纳河上的一个小岛，这里是"一个生活场所，一个非凡与平凡碰撞的地方"。"哥伦比亚彩绘巨匠"是建筑师对麦德林现代美术馆扩建项目的描述（88页）。该建筑由许多盒子堆砌而成（这是一种深受喜爱的设计策略，在赫尔佐格和德梅隆的一些作品中也可以看到类似的结构），有一个可以完全开放的立面，将活动延伸到外部空间，将室内外融为一体。搬到北美之后，Allied Works Architecture建筑设计事务所称他们设计的国家音乐中心（62页）"既是表演大厅、录音设备、播音室，

"Yet to argue that culture is today no longer endowed with the relative autonomy it once enjoyed as one level among others in earlier moments of capitalism […] is not necessarily to imply its disappearance or extinction. On the contrary: we must go on to affirm that the dissolution of an autonomous sphere of culture is rather to be imagined in terms of an explosion: a prodigious expansion of culture throughout the social realm, to the point at which everything in our social life – from economic value and state power to practices and to the very structure of the psyche itself – can be said to have become 'cultural' in some original and as yet untheorized sense."[1]
– Fredric Jameson

Flicking through the architects' descriptions of the projects presented in these pages we are confronted with a use of language that unveils a common interest in blurring boundaries. La Seine Music Hall (p.102), a combination of spaces for the production and performance of music that physically extrudes the tip of a small island on the French river, is presented as "a place for life, where the extraordinary meets the everyday". "A Colombian Piranesi" is how the architects describe the Medellin Modern Art Museum expansion (p.88). Composed through the piling up of boxes (a beloved strategy for similar structures that finds peers in some works of Herzog and de

也是现场音乐场馆和博物馆"。该建筑被划分为两个部分,位于不同的街区,由一座过街大桥重新连接。这种空中连接强调了当代城市与文化活动之间的分离,文化活动占据了一个室内空间,目的是将活动限制在四堵墙内。

如果任何项目描述都具有蛊惑人心的性质的话,那么可能需要将同时满足所有要求的意愿归因于这些项目概要的性质。你很难在任何严格的分类中找到它们,仅仅把它们称为博物馆、剧院或音乐厅,似乎还不足以把握它们的范围。当然,功能分布的模糊有着悠久的历史。SANAA将鹤冈文化馆(74页)定义为"作为该地区文化和艺术活动基地的多功能文化馆",这只是一系列类似声明中的最新一项。这些建筑都属于同一类建筑产品,特别是在20世纪,它们都已经接受了文化建筑的一些普遍特征。可以说,与这一决定相关的是希望最大限度地提高被公众认可的机会。倡导文化是一种获得共识的策略;事实上,这个术语缓和了新建筑与其城市的实际环境和社会环境之间的紧张关系。但是,文化到底是什么呢?

随着时间的推移,我们对文化形成了一种奇怪的定义,就像它宣称自己是一个开放的俱乐部一样——正如弗雷德里克·詹姆逊所言,一切皆为文化——但它同样隐藏了对授予会员资格的一种非常挑剔的态度。艺术和知识实践是普遍获得认可的文化实例,因此,人们很容易同意相关的建筑应该合法地贴上文化的标签。雷蒙德·威廉姆斯清楚地阐述了这一点:"文化是音乐、文学、绘画和雕塑、戏剧和电影……有时还包括哲学、学术和历史。"[2]然而,威廉姆斯对"文化"一词的入侵并没有提出一个含义明确的定义。相反,这位威尔士理论家却忍不住将其当成"英语中最复杂的两三个单词之一"加以介绍。[3]阅读他的文章《关键词》,可以从中发现各种联系,特别是那些被建筑词汇(如文化、文明和城市)广泛使用(而且经常被滥用)的术语之间的联系。

Meuron), the building has one elevation that can be completely opened to extend activities towards the outside space, blending interior and exterior. Moving to North America, Allied Works Architecture talks of their National Music Centre (p.62) as being "at once a performance hall, recording facility, broadcast studio, live music venue and museum". The building is split in two parts located on separate blocks that are reconnected by a bridge overpassing a street. Such aerial connection stresses the disjunction between contemporary city and cultural activities, with the latter populating an interior space that aims to allow for congestion within its four walls.
If any project description has a demagogic character, the willingness to be everything and all at once needs perhaps to be ascribed to the nature of the brief of these projects. Difficult to locate within any strict category, it is as if calling them just museums, or theatres, or music halls, would not suffice to grasp their scope. To be sure, the blurring of programs has a long history. SANAA's definition of Tsuruoka Cultural Hall (p.74) as a "multipurpose hall that serves as a base for cultural and artistic activities in the region" is only the latest in a list of similar statements. These buildings share their belonging to a catalogue of architectural products that, in particular during the 20th century, have accepted taking on the somewhat generic identity of cultural buildings. Arguably, linked to this decision is the hope to maximize chances of being accepted by the public. Advocating culture is tactical to gain consensus; the term, indeed, softens the tensions that always occur between a new piece of architecture and its urban physical and social context. But what is culture, after all?
Over time we have come to a bizarre definition of culture that as much as it declares to be an open club – everything is cultural, as Fredric Jameson stated – it similarly conceals a very selective attitude towards granting membership. Artistic and intellectual practices are commonly accepted instances of culture; hence there is easy agreement that the related architectures should legitimately be labeled cultural. Raymond Williams stated this point clearly: "culture is music, literature, painting and sculpture, theatre and film […] sometimes with the addition of philosophy, scholarship, history."[2] Yet, Williams' incursion into the word culture did not propose a univocal definition. On the contrary, the Welsh theorist could not resist introducing it as "one of the two or three most complicated words in the English language"[3]. When reading his essays *Keywords*, links start to emerge, in particular

威廉姆斯认为，艺术实践代表了20世纪末文化的三重含义之一。其他两种含义则与一个群体、时期或地理区域的特定生活方式以及更加普遍的人类发展过程相互关联。"艺术"最初源自多语义的拉丁文单词colere——同时表示居住、崇拜和照料，它作为一个独立的名词在几个世纪的人类发展过程中被物化了。因此，文化既可以用来表示一种潮流或艺术创作风格，也可以用来表示西方主流文化的存在（比如说，与嬉皮士的反传统文化相对立），或者用来描述已经超越了原始野蛮状态的科技先进的社会。

后者的观点是，文化是一种发展过程，这就标志着与另一个长期以来被用作同义词或反义词的术语的重叠。这个词就是"文明"，这个词与"城市"有着相同的拉丁词源：civis和civitas。但是，civitas的原意并不是指一个有形的物体，而是指属于一个群体（譬如公民群体）的物质条件和象征条件，随着时间的推移，城市也被物化了，其经历与文化类似。众所周知，在威廉姆斯写这篇文章的时候，"城市"这个词似乎已经不能概括人类生活的特定条件了。事实上，到了20世纪70年代初，刻在"城市"一词上的物质和社会纽带的丧失在"城市的胜利"中得到了体现，对此，没有比亨利·列斐伏尔更明确的主张了："社会已完全城市化了"[4]。

通过这种词源学的入侵，我们面临了反对的情况，文化在其中扮演一种积极的角色。随着城市化成为城市的克星，文化有可能成为抵抗文明的工具。至少从18世纪开始是这样的，当时文明开始被用作现代性的核心，并逐渐从与城市共享一个作为有限现象的根源，转向适应城市化的宏大抱负，并为日益壮大的政府政权服务，于是，人们开始有意识地利用文化来抵制这种野心。

刘易斯·芒福德的《城市文化》（1938年）[5]是城市、文明、城市化和文化共享同一擂台的最著名的例子。芒福德是著名的城市化批评家，他详细阐述了城市和建筑规模之间的联系，这与文化在现代性中的明显退化有关。谈到"文化购物"，他认为大型的大都会博

between terms that are widely appropriated – and often misappropriated - by architectural vocabularies, such as culture, civilization, and city.

Williams claimed that artistic practices represented one of three meanings attributed to culture in the late 20th century. Other two connotations related the term reciprocally to the specific way of life of a group, period, or geographical region, and to a more general process of human development. What had originally derived from the multi-semantic Latin term colere – concomitantly signifying to inhabit, to worship and to tend – had been objectified across centuries of human development as an independent noun. As a result, culture had become as much valid to indicate a stream or style of artistic production as to signal the existence of mainstream Western culture (opposed, say, to Hippy counter-culture), or to describe advanced societies that had managed to go beyond a primordial state of barbarism.

The latter aspect – culture intended as a development process – signals an overlap with another term that has for long been alternatively used as synonym or antonym. This word is civilization, which shares the same Latin etymologic roots – civis and civitas – as the word city. But whereas the original meaning of civitas did not relate to a physical object but to the material and symbolic conditions of belonging to a group – a body of citizens – overtime city was subjected to an objectification similar to that experienced by culture. It is known that already by the time of Williams' writing the term city no longer appeared capable of encapsulating a specific condition of human living. By the early 1970s, in fact, the loss of a physical and social bond that was inscribed in the word city had found its representation in the triumph of the urban, about which there is no more explicit a claim than Henry Lefebvre's: "Society has been completely urbanized"[4].

Through such etymological incursion we are confronted with oppositions within which culture is offered an activist role. With urbanization acting as nemesis of city, culture has potentially become a tool of resistance against civilization. This is true at least since the 18th century, when civilization started being used as flagship of modernity and progressively moved from sharing a root with city as a bounded phenomenon, to tuning with urbanization's expansive ambitions at the service of a growing capitalist regime. Culture thus started to be purposely used

物馆相当于百货商店：它们是同一枚硬币的两面，都已成为所谓文明社会的典范现代空间，完全致力于积累欲望。芒福德谴责大型博物馆是权力积累和炫耀鼓吹的场所，这呼应了现代早期先锋派对博物馆更全面的宣战。菲利波·马里内蒂曾在1909年的《未来主义宣言》中写道："我们要拆除各种各样的博物馆、图书馆和院校。"

然而，历史事实证明，一个旧制度的文化设施（博物馆、大学、学校以及图书馆等）是很难真正被摧毁的。很快，人们就清楚地看到，实现这一目标的唯一可能就是从内部采取解散战略。这种操作必然隐含着矛盾，如勒·柯布西耶1939年设计的"可生长美术馆"。对指控的回应转移到了承继自启蒙运动的博物馆，它被当成一个"骗子"，因为"它并没有讲述完整的故事"，勒·柯布西耶的博物馆与芒福德的批评积累形成了尖锐的矛盾——考虑到这两个人物彼此是竞争对手的状态，这一点也不足为奇。该项目认为，保留这个文化机构的意义的唯一方法就是让它成为一切事物的储存库：一个永远在增长的文化机器，在这里，文化的目的是无限的积累，而不是由中央机构强加的、自上而下的人为管理和歧视做法。

将近四十年过去了，蓬皮杜中心承载着1968年的精神，作为一个生产和消费场所，它的诞生最终体现了文化中心的诞生，旨在赋予个体自由，其目标与经常被引用的未实现的先例类似：塞德里克·普莱斯的游乐宫。在建筑史上，蓬皮杜从根本上标志着从高度现代主义到后现代主义风格的过渡，最典型的例子就是弗雷德里克·詹姆逊在开篇语中所认定的文化自主性的丧失。这一巴黎建筑奇迹实际展示了文化领域及其"幽灵一般而又理想化的存在，无论好坏，都超越了现实世界，它把现实世界的镜像以各种形式投射回去，这些形式有的在证明相似性的合理之处，有的又在控诉论辩式的讽刺或乌托邦式的痛苦"。[6]

to resist such ambitions.

A most notable example where city, civilization, urbanization, and culture shared the same boxing ring is Lewis Mumford's The Culture of Cities (1938)[5]. A famous critic of urbanization, Mumford elaborated on the connections between the urban and building scales in relation to a visible degradation of culture in modernity. Speaking of "culture-shopping", he identified the large metropolitan museums as the counterpart to the department stores: the two sides of the same coin, they had become the paradigmatic modern spaces of allegedly civilized societies solely devoted to a desire for accumulation. Mumford's condemnation of the large museum as a place of accumulation and ostentation of power echoed a more general declaration of war to museums by the early modern avant-gardes. The bluntest accusatory words had been uttered by Filippo Marinetti who in the Futurist Manifesto of 1909 had written: "We want to demolish museums, libraries, and academies of every kind."

History tells us, however, that the cultural apparatus of an Ancient Regime (museums, universities, schools, libraries, etc.) proved hard to really be demolished. Soon it became clear that the only possibility towards such goal was a dissolution strategy from the inside. Such operation necessarily implied contradiction, as exemplified by Le Corbusier's Musée à croissance illimitée of 1939. A response to an accusation moved to the museum inherited from the Enlightenment as being a "liar" because "it does not tell the whole story", Le Corbusier's museum was in stark contradiction to Mumford's criticism of accumulation – something that comes as no surprise, considering the status of mutual nemeses of the two figures. The project argued that the only way to retain meaningfulness to this cultural institution was by making it a repository of everything: a perpetually growing machine for culture, where culture was intended as unlimited accumulation opposed to the artificial top-down curatorial, discriminatory practice imposed by centralized institutions.

Almost forty years on, and charged with a 1968 ethos, the Center Pompidou finally embodied the birth of the cultural center as a place of production and consumption meant to empower individual freedom, with a similar goal as its often cited unrealized precedent: Cedric Price's Fun Palace. For the history of architecture, the Pompidou fundamentally signaled the passage from High Modern to Post-Modern by exemplifying the loss of autonomy

1. Fredric Jameson, *Postmodernism, or the Cultural Logic of Late Capitalism*, New Left Review, no. 146 (July–August), 1984, p.87.
2. Raymond Williams, *Keywords: A Vocabulary of Culture and Society*, Fontana Press, London, 1976, p.90.
3. Ibid. p.87.
4. Henry Lefebvre, *The Urban Revolution*, University of Minnesota Press, Mineapolis, 2003 (first published as La Révolution Urbaine, 1970), p.1.
5. Lewis Mumford, *The Culture of Cities*, Harcourt, Brace and Co, New York, 1938.
6. Fredric Jameson, cit. p.87.

蓬皮杜中心的存在模糊了合法性和对现状的争论，标志着一个稍纵即逝的时刻。其追随者们则是一诞生就被烧得焦头烂额，尽管这些后来的同类建筑经常声称拥有同样的雄心壮志，但却不得不面对变化的新自由主义政治和经济制度，在这种情况下，（随着建筑设计的复杂性日益提高）就自由生长、细分和封闭的毛细作用力，以及建筑环境奇观化的相关现象等方面进行争论变得愈发困难。然而，那些仍然被贴上文化标签的项目，至少在其公开的意图中，仍然保持着最初的抵抗目标，在本书所挑选的项目中情况可能也是如此，其模糊的目标揭示了一种更普遍的态度，即对日益敌对的城市局势做出反应。一旦城市不复存在，城市化就变得无处不在，不论是好是坏，对某种形式的社会联系的责任就转移到了建筑上。机会来了，如果不是公开辩论，至少也是更微妙的评论。

特别是其中一个项目嵌入了对文化的评论，它认为文化是一种反抗持续城市化的逻辑及与之相关的后期资本主义社会价值观商品化的行为。吉列尔莫·巴斯克斯·孔苏埃格拉在塞维利亚的卡伊莎中心（46页）用一个最简单的策略标记了领地，这个策略是让公共空间在地下不间断地流动。树冠的复杂形状表明，必须与西萨·佩里设计的塞维利亚塔及相关地标性建筑共存，这让人感到有些尴尬。剖面图展示了建筑的竖井，目的是让自然光进入实际建筑物，该建筑试图容纳所需的文化功能，最终，它又不可避免地成为一处地下购物中心，为市区的大型重建项目提供支援。在这种策略下，文化的标签表达了一种将城市从当代商品化的城市状态中拯救出来的戏剧感：其紧迫性与日俱增。

of culture diagnosed by Fredric Jameson in the opening quotation. The Parisian marvel gave physical representation to the cultural realm and "its ghostly, yet Utopian, existence, for good or ill above the practical world of the existent, whose mirror image it throws back in forms which vary from the legitimations of flattering resemblance to the contestatory indictments of critical satire or Utopian pain".[6]
Blurring legitimation and contestation of the status quo, the Pompidou marked a fleeting moment. Burned immediately after their birth, its followers often claimed similar ambitions but had to confront the changed political and economic regime of neo-liberalism, which has made contestation an increasingly harder task against the capillary forces of free growth, subdivision, enclosure and the related phenomena of spectacularization of the built environment (with the complicity of architecture). Nevertheless, the projects that still claim the label of culture maintain, at least in their declared intentions, an original objective of resistance, and this might hold true also for the projects selected here, whose blurring objectives unveil a more general attitude of responding to increasingly hostile urban situations. Once the city is no more, and the urban is everywhere, for better or worse the responsibility towards some form of social bond shifts to the building. And the opportunity arises, if not for an open polemic, at least for more subtle commentary.
One of the projects, in particular, embeds a comment about culture as an act of resistance against the logics of continuous urbanization and the related commodification of values of late-capitalist society. Guillermo Vàzquez Consuegra's CaixaForum in Seville (p.46) marks territory with the simplest strategy of a canopy letting public space flow uninterrupted underneath. The canopy's complex shape signals some embarassment about having to co-inhabit with the landmark architecture of César Pelli's Sevilla Tower and related buildings. Sectional drawings unveil a vertical shaft intended to let natural light into the actual building, which tries to accommodate the required cultural functions inside what is, ultimately, one more inevitable underground shopping mall supporting a large redevelopment project for an urban area. Attached to this tactic, the label of culture expresses the drama of redeeming the city from the contemporary commodified urban condition: something whose urgency grows stronger every day.

塞尔维亚卡伊莎文化中心
CaixaForum Sevilla
Guillermo Vázquez Consuegra

西班牙储蓄银行是西班牙三大私人银行之一。它侧重于公共社会问题、艺术和科学、研究和教育方面。自2002年以来，该银行通过在加泰罗尼亚和马德里开设卡伊莎文化中心，在西班牙的各大城市均进行了投资。

2009年，塞维利亚的建筑师吉列尔莫·巴斯克斯·孔苏埃格拉赢得了一场设计竞赛，在建于15世纪的塞维利亚皇家造船厂打造第八座卡伊莎文化中心，这是继马德里和巴塞罗那之后规模排行第三的卡伊莎中心。该文化中心占据了塞维利亚地下停车场的一部分，该停车场由塞维利亚设计师西萨·佩里设计，高180m。托雷·塞维利亚综合体位于1992年塞维利亚世博会的旧址上，穿过从历史中心流出的瓜达尔基维尔河支流。世博会结束后，该地区被遗弃，但人们希望卡伊莎文化中心和佩里的综合体将促进该地区的重建。塞维利亚卡伊莎文化中心占地8100m²，包括两个展览厅，一个礼堂，两个多功能厅，一个儿童活动区，一个商店和一个带室外露台的咖啡馆餐厅。它的创建是为了举办社会和文化活动，从艺术展览、科学主题到摄影和视觉艺术，涵盖甚广。它的目的是宣传知识，把艺术、音乐和人文学科带入社会。该项目需要改造托雷·塞维利亚项目现有的裙楼，并在主入口加建一个由稳定泡沫铝材料（SAF）覆盖的广场，采光天窗可将日光引入地下。

入口雨篷与综合体另一端的椭圆塔的雨篷相匹配，虽然体积较小，但也具有相应的代表性。巴斯克斯·孔苏埃格拉用充满魅力、直观的拱顶和曲线塑造了雨篷的表面，提供了柔和而又不断变化的反射光。这个雨篷的外形让人联想到它及其下方空间所不具备的品质：这

是一个挖开的、有雕刻感的实体,呈厚重的拱形,可能也会让我们想起以前皇家造船厂的拱顶。游客通过位于摩天大楼周围的公园平台往下走进入室内。门廊、门道用来指示公园的入口,或者考虑到在构图和独特性方面的价值,也可能是一个小教堂。然而在这种情况下,更多的元素得到了使用;这座公园作为通往地下世界的象征性入口,不但成了一个入口,而且可能也成了某种在时间上更为久远的事物,一种全新的体验:21世纪的史前墓石碑坊。

宽敞的、两层高的地下大厅充满了自然光线,光线透过雨篷的一根实体墩柱照射下来,其弯曲的玻璃表面在室内采用较轻的多孔SAF面板进行遮蔽,创造了可以在地板和墙壁上移动的、斑驳的光线。在往下走到大厅的过程中,第一段机械楼梯呈弯曲状,弯曲角度与结构网格相关,使空间中的移动路线更加灵活。

主厅的总体氛围让人联想起一个优雅的地铁站,无论是金属表面和用菱镁矿石铺就的闪闪发光的地板、外观时髦的沙发和家具,还是未完工的天花板上铺设的连续LED灯管,偶尔会有广场上花盆的混凝土盒子伸出来。较小的自然采光大堂的焦点是一条宏伟的旋转楼梯,楼梯一直延伸到餐厅,餐厅拥有一个可以俯瞰广场的架高平台。挂在弯曲钢轨上的厚重窗帘悬挂在天花板下方,给人一种失重的感觉。向上走到一层,游客将再次注意到来自餐厅和露台的自然光。

一侧画廊的外墙采用了青铜色的金属板当饰面,礼堂的墙壁是蓝绿色的,它会随着人们在周围走动而改变色调。

La Caixa is one of Spain's three largest private banks. It focuses on public social issues, the arts and sciences, research and education. Since 2002, the bank has invested in Spain's cities by opening CaixaForum cultural centers in Catalonia and in Madrid.

In 2009, Sevillano architect Guillermo Vazquez Consuegra won a competition to install the eighth CaixaForum, the third largest after Madrid's and Barcelona's, at the 15th century Royal Shipyards in Seville. The Forum occupies part of the underground parking garage of the 180m Torre Sevilla, designed by César Pelli. The Torre Seville complex is located on the former grounds of the 1992 Seville Expo, across a branch of the Guadalquivir River from the historic center. After the Expo, the area is abandoned, but the hope is that the CaixaForum and the Pelli complex will help revive it.

CaixaForum Seville covers 8,100m², consisting of two exhibition halls, an auditorium, two multi-use rooms, a kids zone for children's activities, a shop, and a café-restaurant with an outdoor terrace. It is created to host social and cultural events, from art exhibitions, scientific subjects, photography and visual arts. Its purpose is to promote knowledge, and bring art, music and humanities to society. The project requires adapting the existing Pódium building that is part of the Torre Seville – and adding a plaza covered by Stabilized Aluminum Foam (SAF) to the main entrance, with a skylight to let daylight into the underground levels.

The entry canopy matches that of the elliptical tower at the other end of the complex, and it claims a correspond-

项目名称：Caixaforum Sevilla / 地点：Isla de la Cartuja, Sevilla, Spain
建筑师：Guillermo Vázquez Consuegra
技术建筑师：Marcos Vázquez Consuegra - project and direction of works;
Ignacio Gonzalez - direction of works
合作方：Juan José Baena - coordinator; Eduardo Melero; Alberto Brunello;
Martina Pozzi; Álvaro Luna; Patricia Fraile - furniture
结构工程师：Edartec Consultores, S.L / 设备：Ingenieros-JG, S.L
声学：Higini Arau / 风景咨询：GD Consulting
金属工程咨询：Jorge Vázquez Consuegra
项目经理：Carlos Coronado, Maria Dolores Casado e Idom
标识：Estrada Design
模型：G-Metal and Estudio Vázquez Consuegra / 承包商：Dragados S.A.
客户：Fundación Caixa D'Estalvis i Pensions Barcelona "La Caixa"
用地面积：8,000m² / 建筑面积：10,500m² / 总建筑面积：8,930m²
结构：fit inside an existing reinforced concrete structure; new elements(canopy,
self-standing walls, etc) - steel structure
材料：aluminium foam panel, aluminium, glass, iron sheets,
anodised aluminium panels, magnesite flooring
造价：9.628.500 € / 设计时间：2015 / 竣工时间：2017
摄影师：
©Duccio Malagamba (courtesy of the architect) - p.46~47, p.48~49, p.50, p.57, p.58, p.60
©Jesus Granada (courtesy of the architect) - p.52, p.53, p.59, p.61

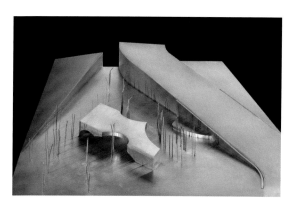

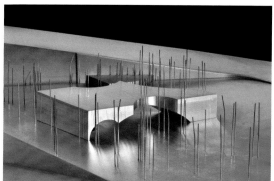

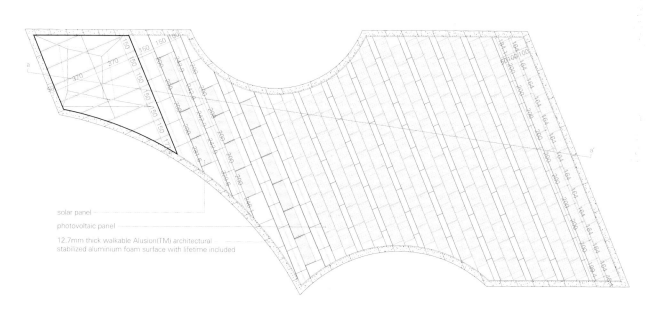

详图1——太阳能光伏盖板
detail 1 _ solar and photovoltaic cover

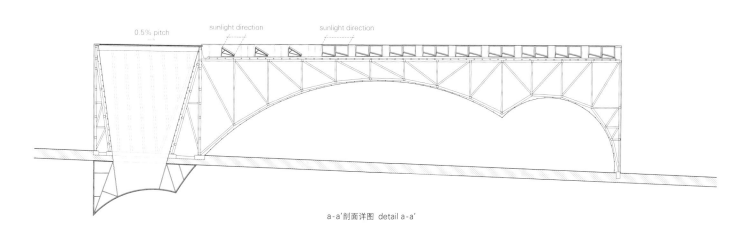

a-a'剖面详图 detail a-a'

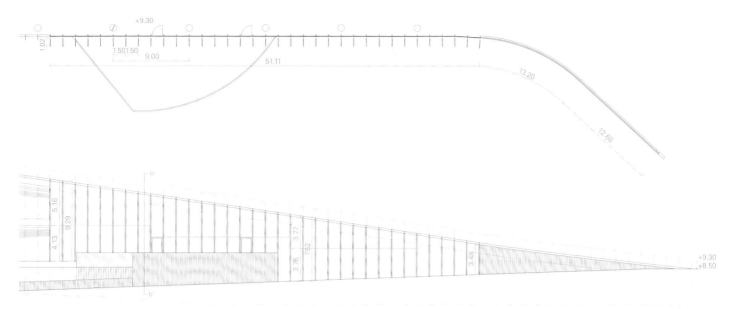

详图2 detail 2

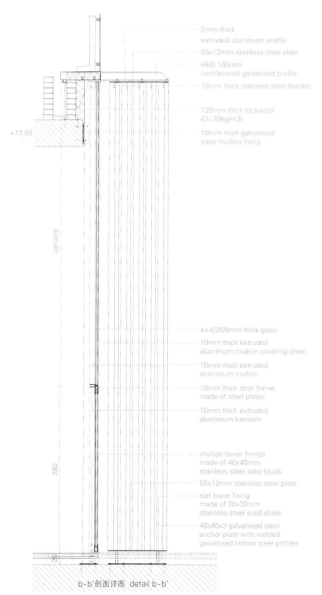

2mm thick extruded aluminium profile
55x12mm stainless steel plate
HEB 140mm cantilevered galvanised profile
10mm thick stainless steel bracket
120mm thick rockwool (D=70kg/m3)
10mm thick galvanised steel mullion fixing

4+4/20/8mm thick glass
10mm thick extruded aluminium mullion covering sheet
10mm thick extruded aluminium mullion
10mm thick door frame made of steel plates
10mm thick extruded aluminium transom

mullion lower fixings made of 40x40mm stainless steel solid studs
55x12mm stainless steel plate
slat lower fixing made of 30x30mm stainless steel solid studs
40x40x3 galvanised steel anchor plate with welded galvanised hollow steel profiles

b-b'剖面详图 detail b-b'

ing representative role despite its diminutive size. Vázquez Consuegra has molded its surfaces with seductive, intuitively sketched vaults and curves, which offer soft, changing registers of reflected light. The canopy's forms evoke qualities that neither it nor the spaces below it possess: an excavated, carved solid, vaulted and heavy, references that may also remind us of the vaults of the old Royal Shipyards. Visitors descend to the interior via a platform situated in the park surrounding the skyscraper. Porches, gateways or folíes are used to indicate the entrance to a park, or perhaps a tempietto on account of its value in terms of composition and uniqueness. In this case, however, something more was used; the park's status as a symbolic entryway to the underground world makes it a propylaeum, or perhaps something far more distant in time, a new experience: a 21st-century dolmen.

The spacious, double-height underground lobby is dominated by the natural light that drops through the canopy's one solid pier. Its curvy glazed faces are screened inside with lighter, porous SAF panels, creating planes of moving, dappled light across the floors and walls. In the descent to the lobby, the first run of mechanical stairs is cranked at an angle in relation to the structural grid, giving movement through the space another dynamic vector.

The general atmosphere of the main lobby recalls a classy metro station, with its metallic surfaces and gleaming mag-

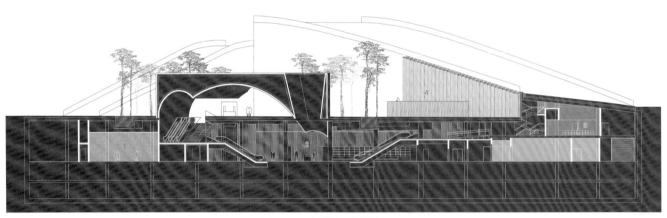

1. 展室 2. 主厅 3. 多功能室 4. 咖啡馆/餐厅
1. exhibition room 2. main lobby 3. multi-purpose room 4. cafe/restaurant
A-A'剖面图 section A-A'

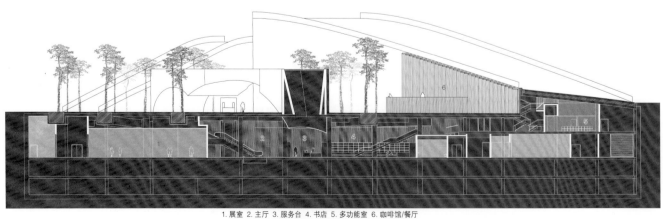

1. 展室 2. 主厅 3. 服务台 4. 书店 5. 多功能室 6. 咖啡馆/餐厅
1. exhibition room 2. main lobby 3. information desk 4. bookshop 5. multi-purpose room 6. cafe/restaurant
B-B'剖面图 section B-B'

nesite floors, its sleek sofas and furnishings, and its continuous LED lighting tubes on the unfinished ceiling, where the occasional concrete box of the plaza's planters pokes down. The smaller, naturally-illuminated lobby is dominated by a magnificent switchback stair that rises to the cafeteria, which boasts a raised deck overlooking the plaza. Heavy drapes hung from curved steel rails suspended from the ceiling create a feeling of weightlessness. Moving up the ground floor, visitors will again notice the natural light that flows from the restaurant and terrace.

The exterior walls of the galleries on one side are finished in bronzed metal panels, and the auditorium wall in a greenish-blue changes the hue as one moves around it.

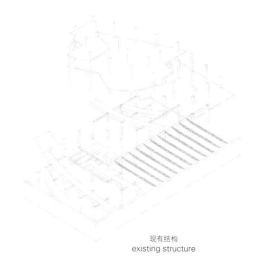

现有结构
existing structure

新结构
new structure

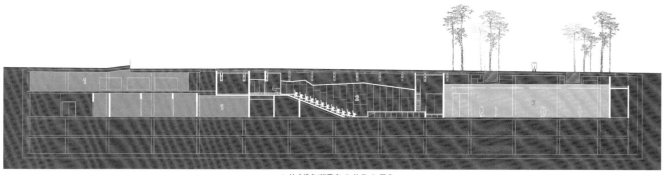

1. 技术设备/储藏室 2. 礼堂 3. 展室
1. technical/storage rooms 2. auditorium 3. exhibition room
C-C'剖面图 section C-C'

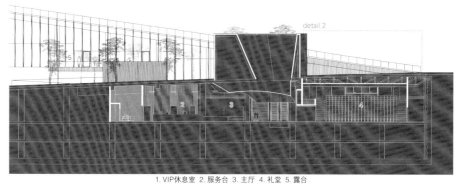

1. VIP休息室 2. 服务台 3. 主厅 4. 礼堂 5. 露台
1. VIP lounge 2. information desk 3. main lobby 4. auditorium 5. terrace
D-D'剖面图 section D-D'

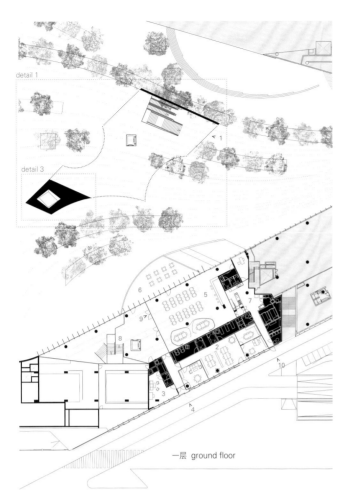

1. 卡伊莎文化中心主入口	1. Caixaforum main entry
2. 管理区	2. management area
3. VIP休息室	3. VIP lounge
4. 管理和VIP通道	4. administration and VIP's access
5. 咖啡馆/餐厅	5. cafe/restaurant
6. 露台	6. terrace
7. 餐厅厨房	7. restaurant kitchen
8. 下行楼梯	8. stairs down
9. 咖啡馆/餐厅入口	9. entry to the cafe/restaurant
10. 厨房入口	10. kitchen entry
11. 礼堂	11. auditorium
12. 一号多功能室	12. multi-purpose room 1
13. 二号多功能室	13. multi-purpose room 2
14. 儿童实验室	14. children lab
15. 主入口,向上通往广场的楼梯	15. main entrance, stairs up to square
16. 向下通往地下大厅的楼梯	16. stairs down to underground lobby
17. 礼堂大厅	17. auditorium foyer
18. 向上通往咖啡馆/餐厅的楼梯	18. stairs up to cafe/restaurant
19. 多功能室门厅	19. multipurpose room foyer
20. 卫生间	20. toilets
21. 向上通往一层的次要通道	21. secondary access up to ground floor level
22. 技术设备/储藏室	22. technical/storage rooms
23. 装卸货入口坡道	23. loading/unloading entrance ramp
24. 一号展室	24. exhibition room 1
25. 二号展室	25. exhibition room 2
26. 服务台	26. information desk
27. 书店	27. bookshop
28. 主厅	28. main lobby
29. 向上通往带顶广场的主入口	29. main entrance up to covered square
30. 上行楼梯	30. stairs up
31. 储物柜	31. lockers
32. 更衣室	32. dressing room
33. 安检口	33. security checkpoint
34. 书店储藏室	34. bookshop storage room
35. 设备/储藏室	35. equipment/storage rooms
36. 设备走廊	36. services corridor
37. 机械房	37. mechanical room

一层 ground floor

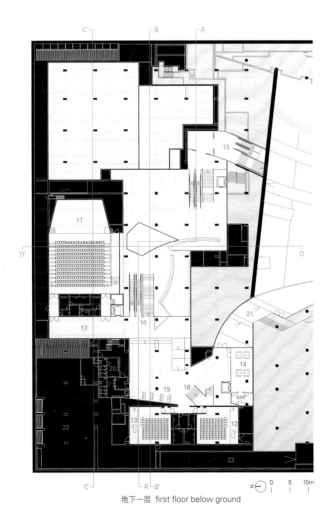

地下一层 first floor below ground

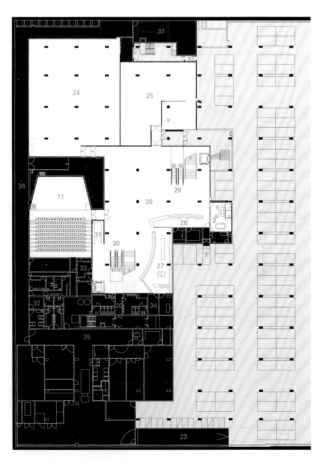

地下二层 second floor below ground

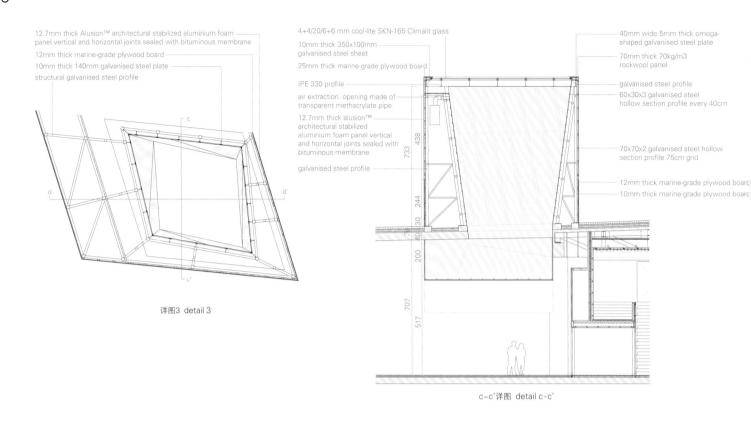

详图3 detail 3

c-c'详图 detail c-c'

12.7mm thick Alusion™ architectural stabilized aluminium foam panel vertical and horizontal joints sealed with bituminous membrane
12mm thick marine-grade plywood board
10mm thick 140mm galvanised steel plate structural galvanised steel profile

4+4/20/6+6 mm cool-lite SKN-165 Climalit glass
10mm thick 350x100mm galvanised steel sheet
25mm thick marine-grade plywood board
IPE 330 profile
air extraction. opening made of transparent methacrylate pipe
12.7mm thick alusion™ architectural stabilized aluminium foam panel vertical and horizontal joints sealed with bituminous membrane
galvanised steel profile

40mm wide 5mm thick omega-shaped galvanised steel plate
70mm thick 70kg/m3 rockwool panel
galvanised steel profile
60x30x3 galvanised steel hollow section profile every 40cm
70x70x2 galvanised steel hollow section profile 75cm grid
12mm thick marine-grade plywood board
10mm thick marine-grade plywood board

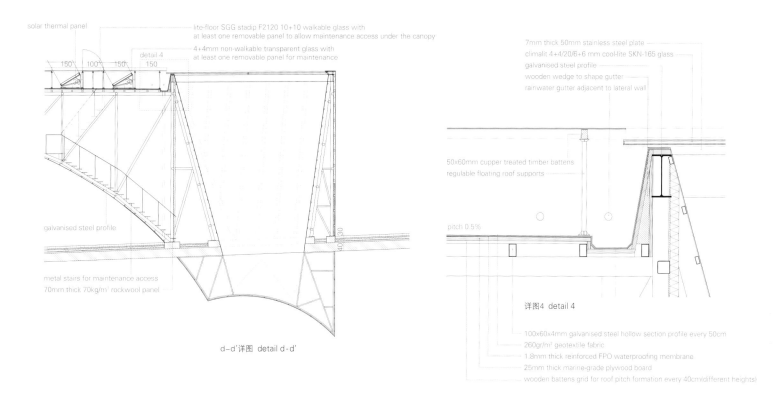

d-d'详图 detail d-d'

详图4 detail 4

作为抵抗方式的文化 Culture for Resistance

贝尔工作室——加拿大国家音乐中心所在地
Studio Bell, Home of the National Music Center

Allied Works Architecture

早期的概念模型由铜管乐器的
部件构成。
An early concept model incorporates
segments of brass instruments.

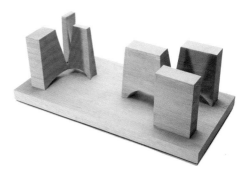

后来的模型将博物馆的概念展示为
一系列容器。
A later model displays the concept of
the museum as a series of containers.

2009年，经过一场国际设计竞赛和层层考察，Allied Works Architecture建筑事务所获得了加拿大国家音乐中心（NMC）新总部——贝尔工作室的设计委托。该建筑于2016年10月完工，是北美首个此类设施，它既是表演大厅、录音设施、广播演播室、现场音乐场馆和博物馆，也是Allied Works Architecture迄今为止最雄心勃勃的项目。这座先进的文化中心以互动展览、教育和表演空间为特色，并整合附近建于1905年的爱德华国王酒店（传说中的蓝调俱乐部旧址），给酒店重新带来了活力。

贝尔工作室由九座互相连接的塔楼组成，表面覆盖定制的釉面陶土瓷砖，在其微妙的曲线设计中参考了声学容器。该建筑由两个主要的结构系统组成：第一个在一层形成横跨大厅的连锁拱门，第二个悬挂主要的表演空间，并从上方连接塔楼。在贝尔工作室的五层楼中，走道和楼梯将两个系统连接起来，在这里，琉璃瓦的相互作用产生了反射光线和放大音量的效果。

游客通过中央大厅进入贝尔工作室，大厅向上延伸至建筑的五层。大厅的南侧和北侧有两个螺旋楼梯，填补了塔楼之间的空隙。主要表演大厅不但可俯瞰大堂，也可作为建筑的布局和结构中心。有了灵活的座位和可移动的隔声墙，表演大厅可以关闭，进行更亲密的演出活动，也可以面向大厅和交通流线空间完全开放，让乐曲声回荡在整座建筑的各个角落。

展厅或"舞台"分布在贝尔工作室的五个楼层，展示国家音乐中心收藏的亮点。每个舞台都是一个互动、欣赏和表演的地方，在这里，人们可以进一步探索加拿大过去、现在和未来的音乐故事。每两个空间之间都有过渡空间，让游客反思并与建筑重新产生链接，为下一次的相遇做准备。在整座建筑的各个位置，牛仔竞技公园和弓河的美景都一览无余，这让游客可以与周围的城市和景观重新产生联系。

Allied Works Architecture was awarded the commission for the design of the National Music Center's (NMC) new home, Studio Bell, in 2009, following an international design competition and search. Completed in October 2016, the building is the first facility of its kind in North America – at once a performance hall, recording facility, broadcast studio, live music venue and museum – and Allied Works' most ambitious project to date. The state-of-the-art cultural center features interactive exhibition, education, and performance spaces, and incorporates and revitalizes the neighboring historic 1905 King Edward Hotel, the former home to the legendary blues club.

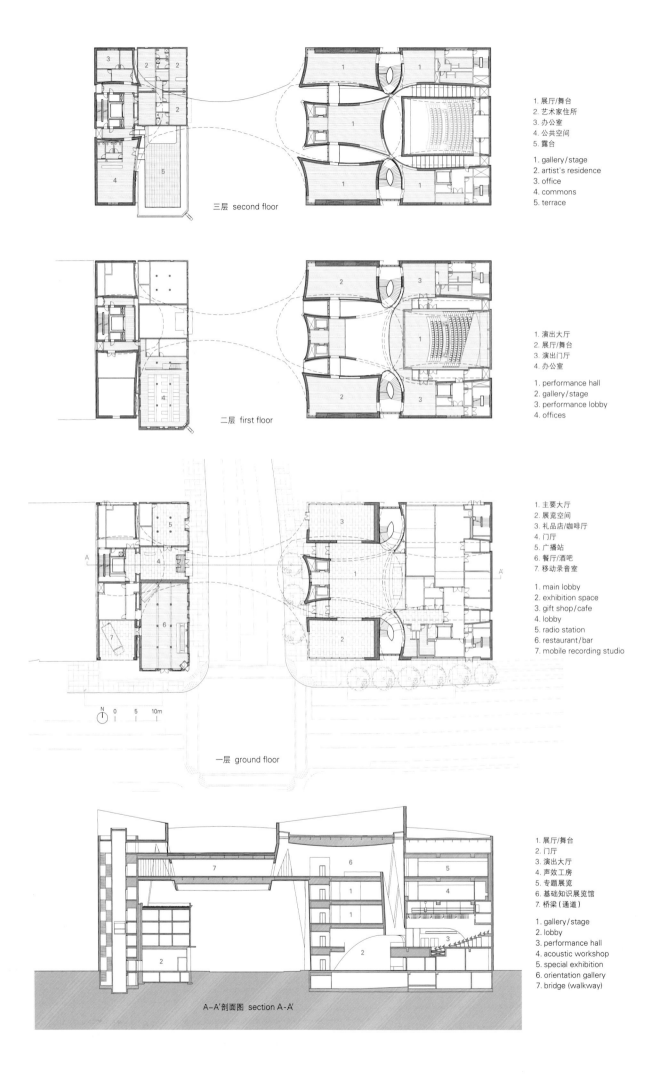

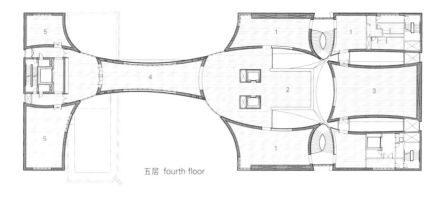

1. 展厅/舞台
2. 基础知识展览馆
3. 专题展览
4. 桥梁（通道）
5. 教室

1. gallery/stage
2. orientation gallery
3. special exhibition
4. bridge (walkway)
5. classroom

五层 fourth floor

1. 展厅/舞台
2. 声效工房
3. 电子工房
4. 远程教育工作室
5. 录音棚

1. gallery/stage
2. acoustic workshop
3. electric workshop
4. distance learning studio
5. recording studio

四层 third floor

Rising in nine interlocking towers, clad in custom-glazed terra cotta tile, Studio Bell references acoustic vessels in its subtly curved design. The building is comprised of two main structural systems – the first forms the interlocking arches that span the lobby on the ground level, the second suspends the primary performance space and bridges the building's towers from above. Walkways and stairs unite the two systems throughout Studio Bell's five stories, where an interplay of glazed tile reflects and amplifies light and sound.

Visitors are welcomed into Studio Bell through a central lobby that opens upward through the building's five floors. Two helical staircases on the north and the south flank the lobby and fill the interstitial space between the towers. The main performance hall overlooks the lobby, and serves as the building's programmatic and structural center. With flexible seating and a movable acoustic wall, the performance hall can be closed for more intimate performances or opened fully to the lobby and circulation spaces to fill the entire building with music.

Exhibition galleries or "stages" are spread across Studio Bell's five floors, showcasing highlights from NMC's collection. Each stage is envisioned as a place for interaction, appreciation and performance, where Canada's music story – past, present and future can be further explored. Between each is a pause of space that allows the visitor to reflect and reconnect to the building and prepare for the next encounter. Sweeping views of Stampede Park and the Bow River throughout the building allow visitors to connect back to the surrounding city and landscape.

项目名称：Studio Bell, Home of the National Music Center
地点：East Village district of Calgary, Alberta, Canada
建筑师：Allied Works Architecture
总建筑师：Brad Cloepfil
总负责人：Kyle Lommen
项目经理：Chelsea Grassinger
项目建筑师：Daniel Richmond, Dan Koch
项目团队：Kyle Caldwell, Paul Bickell, Brock Hinze, Bjorn Nelson, Philip Balsinger
合作建筑师：Kasian Architecture
总承包商：CANA Construction
展览设计：Haley Sharpe Design
剧院设计：Fisher Dachs Associates
结构工程：RJC
电子工程：SMP
机械工程：Stantec
声学&A/V设计：JaffeHolden
标题设计：Royal Tichelaar Makkum
展览内容设计：St. Joseph Media
录音棚设计：Pilchner Schoustal International Inc.
总建筑面积：16,852.61m²
- 14,864.49m² _ new construction
- 2,043.87m² _ exhibition galleries
- 371.61m², 300 seat _ recital hall
材料：custom glazed terra cotta cladding, steel and concrete structure
认证：targeted LEED silver
竣工时间：2016.10
摄影师：©Jeremy Bittermann (courtesy of the architect)

鹤冈文化大厅
Tsuruoka Cultural Hall

SANAA

鹤冈文化大厅作为该地区文化和艺术活动的基地，是一个多用途的建筑。周围场地的特色是拥有丰富的自然环境，而且坐落于城市的文化区域，历史建筑、大学院校和美术馆比比皆是。这座建筑试图扩大当地市民活动的范围，同时合并了旧有的文化馆，这里是学生和当地艺术团体举行文化活动的核心。

作为一个市民在此充当参与者、观察者和演员的"社区大厅"，它被一条走廊环绕着，类似于传统的日本"Saya-do"建筑。走廊每天都对公众开放，公众可以在不同的地方活动，而不需要区分"前面的"或"后面的"空间。无论什么时候举行专业表演，舞台上方和下方的隔断都可以营造私密的后台空间。通过尽量减小中央大厅的进深，拉近了观众和表演者的距离。大厅被建筑师设计成葡萄园风格，声音在整个大厅内部回荡，从任何座位都有一条通往舞台的直接通道。

这座建筑物的外观看起来像许多小屋顶的集合体。每一个小屋顶在朝向建筑四周的方向都会降低，就像路边的单层建筑那么低。建筑师通过这种方式来控制体量，试图使建筑与邻近的历史建筑以及周围的城市景观相协调。

金属板材、抹灰混凝土饰面、弯曲的钢框架的纹理，以及木百叶所具有的自然和温暖的感觉，无不给人一种手工建造建筑的感觉。整座建筑处处都能让人感受到人性的温暖，并与周围的自然产生对话。建筑元素与自然环境相互作用，其表现形式随着时间、光线等自然条件的变化而变化。鹤冈文化大厅将成为这座城市优美风景的一部分，支持当地社区的文化艺术活动。

Tsuruoka Cultural hall is a multipurpose hall that serves as a base for cultural and artistic activities in the region. The surrounding site is characterised by its abundant nature, and is located in a cultural precinct of the city filled with historic buildings, universities and galleries. A building was sought to expand local civic activities, while incorporating the old cultural hall which has been the heart of cultural activities to the students and local arts groups.

As a "hall for the community" where citizens become participants, observers and actors, a large hall was proposed

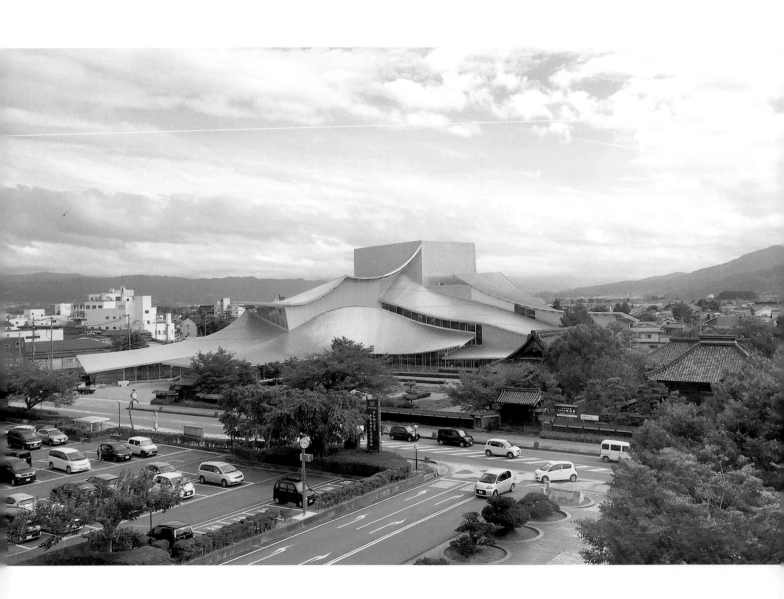

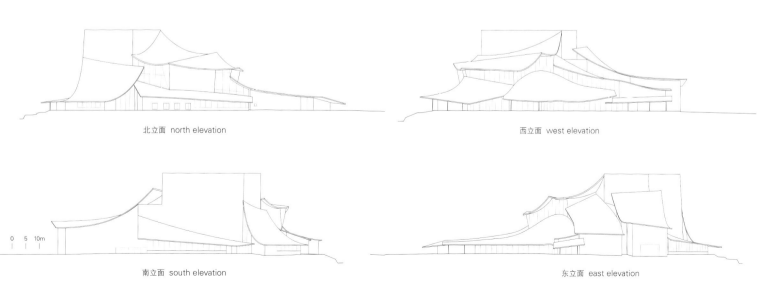

| 北立面 north elevation | 西立面 west elevation |
| 南立面 south elevation | 东立面 east elevation |

项目名称：Tsuruoka Cultural Hal / 地点：Babacho 11-61, Tsuruoka City, Yamagata Prefecture, Japan / 建筑师：SANAA + Shinbo Architects Office + Ishikawa Architects Office
团队：Kazuyo Sejima (Principal), Ryue Nishizawa (Principal), Rikiya Yamamoto (Partner), Satoshi Ikeda, Takuma Yokomae, Kohei Kudo, Hayao Odagiri – SANAA; Takashi Maruyama, Taiji Ota – Shinbo Architects Office; Yuji Ishikawa, Tomohiro Ishikawa – Ishikawa Architects Office / 结构：ARUP / 机械工程师：SOGO CONSULTANTS
剧院设计：Shozo Motosugi / 剧院声效：Nagata Acoustics / 剧院照明：Motoi Hattori / 气候：Takahiro Chiba / 客户：Tsuruoka City

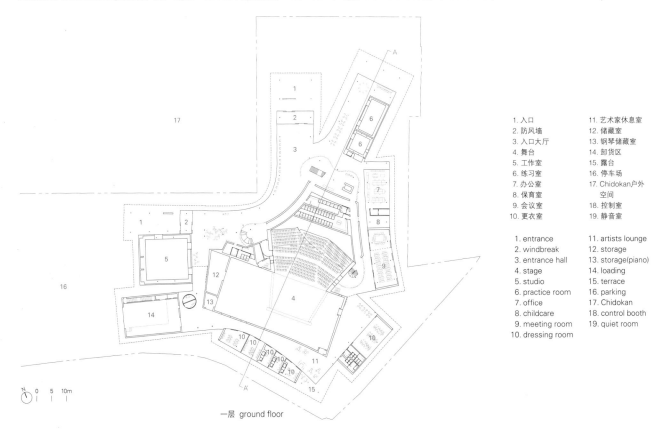

1. 入口	11. 艺术家休息室
2. 防风墙	12. 储藏室
3. 入口大厅	13. 钢琴储藏室
4. 舞台	14. 卸货区
5. 工作室	15. 露台
6. 练习室	16. 停车场
7. 办公室	17. Chidokan户外空间
8. 保育室	18. 控制室
9. 会议室	19. 静音室
10. 更衣室	

1. entrance	11. artists lounge
2. windbreak	12. storage
3. entrance hall	13. storage(piano)
4. stage	14. loading
5. studio	15. terrace
6. practice room	16. parking
7. office	17. Chidokan
8. childcare	18. control booth
9. meeting room	19. quiet room
10. dressing room	

一层 ground floor

施工团队：Takenaka Corporation, Sugawara Corporation, Suzuki Corporation JV / 用途：multipurpose hall / 用地面积：13,096.84m² / 建筑面积：5,756.35m² / 总建筑面积：7,846.12m² / 结构：steel + RC / 基础：RC / 通风：machine ventilation / 供暖、制冷系统：AHU, PAC / 电气系统：cubicle type high voltage receiving equipment / 外部装修：glass, aluminum panel, plated steel sheet / 室内装修：Entrance hall - polished concrete (floor), concrete (wall), wooden louver (ceiling); Main hall - wood flooring (floor), wood panel (wall) / 施工造价：8,486,217,720 JPY / 设计时间：2012.8—2013.11 / 施工时间：2014.10—2017.8 / 开放时间：2018.3
摄影师：courtesy of the SANAA

二层 first floor　　　　　　　　　　　　　屋顶 roof

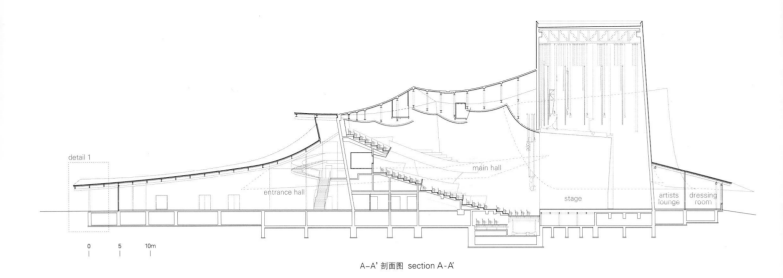

A-A' 剖面图 section A-A'

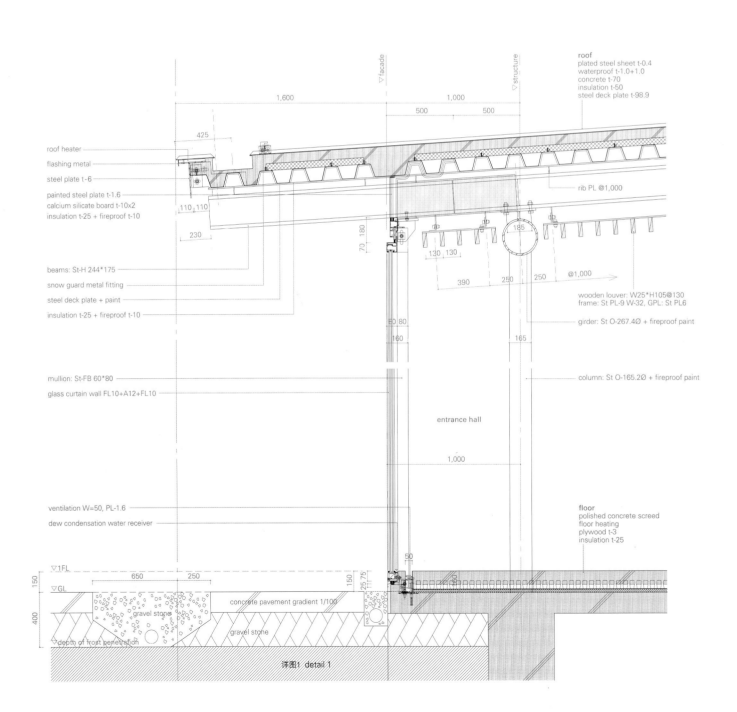

详图1 detail 1

wrapped by a corridor, resembling traditional Japanese "Saya-do" construction. The corridor is open to the public on a daily basis, and can be active in various places without distinguishing "front" or "backhouse" spaces. Whenever a professional performance takes place, partitions above and below the stage can create private back of house spaces. By keeping the depth of the large central hall as small as possible, the audience and performer are brought together with a sense of unity. The hall is designed in a vineyard style, where the sound echoes through the entire hall and there is a direct path to the stage from any seat.

The exterior of the building looks like an assembly of many small roofs. Each small roof becomes lower towards the perimeter of the building, as low as a one-story building along the road. By controlling the volume in this way, the architect seeks to make a building that is in harmony with the adjacent historic building and the surrounding cityscape. Textures of sheet metal, plastered concrete finishes, curved steel framing, and the naturalness and warmth given by the wooden louvres – give a sense of architecture created by human hands. The warmth of humanness is felt throughout the building, and creates a dialogue with the surrounding nature. The architectural elements interact with the natural environment, and their expressions change according to natural conditions such as time and light. The Tsuruoka Cultural Hall will become a part of the city's beautiful landscape, supporting the cultural and artistic activities of the local community.

麦德林现代美术馆扩建
Medellín Modern Art Museum Expansion

51-1 + Ctrl G

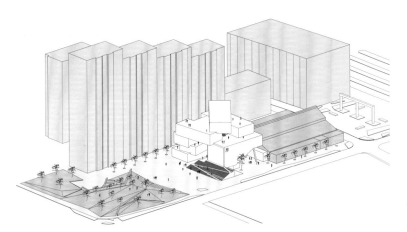
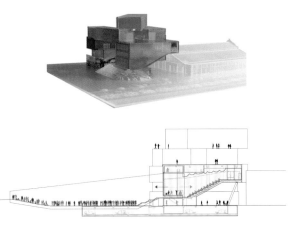

麦德林现代美术馆由一群艺术家于1978年创建,旨在唤起公众对现代和当代艺术的兴趣。

2006年,麦德林将西梅萨钢铁厂的工业用地改造成了一个混合用途的社区,起名为"Ciudad del Rio"(河畔之城)。城市规划考虑拆除所有的工厂,以便在一套严格的方案中设计一座一般的高层建筑,只留下最古老的中央广场——Talleres de Robledo,作为工业历史的唯一见证,在现代美术馆成立30周年之际,它被神话般的Grupo Utopia公司成功翻新成为现代美术馆的新场所。

2009年,MAMM组织了一次国际设计竞赛,以满足世界级博物馆的扩建需求。由西班牙建筑师Federico Soriano担任主席的评审团和参观了展览方案的公众,投票将一等奖授予了来自秘鲁的51-1团队和来自哥伦比亚的Ctrl G团队。

获奖的方案起源于麦德林当地山坡上层叠搭建起来的、非正规的聚居群落模式。街区里的建筑层层叠叠,形成了数以千计的公共空间,人们在这里以最巧妙的方式展现自己的城市风貌。

与其说是建筑,倒不如说是村庄

与其说是建筑,倒不如说是垂直的村庄。通过这种方式,就像山坡上堆叠房屋的非正式隔墙那样,一户人家的屋顶正是楼上邻居的露台,在MAMM的扩建中,所需的功能空间被堆叠起来,形成了由楼梯连接的一连串逐层升高的公共露台。内部是一条艺术空间长廊。室外,Ciudad del Rio的带状公园的垂直延伸一直延伸到五楼:每一个露台都为未来博物馆的改造预留了空间。

11个盒子和一个挑空空间

博物馆共有五层。除了新建的艺术画廊之外,它还拥有实验室、储藏室、办公室、商店、咖啡馆和剧院。这些区域需要封闭、密封,并具有特殊的光线和几何条件,换言之,它们就像盒子一样。每个盒子都由预制混凝土板封闭,这些预制混凝土板经过穿孔、切割和雕刻,呈现出不同的颜色、光线和纹理。11个大小和高度不同的盒子,在精细布局的位置上旋转并堆叠在一起。这个介于两者之间的空间是一个功能混合的地方,在这里,人与人之间的关系日渐丰富,也进一步增强了与景观的联系。

一体化

从中央广场延伸到建筑物的中心,在51-1和Ctrl G负责扩建的中殿内部,形成了博物馆的新主入口:这是一个促成新旧空间相连的巨大空间,它的目的是为了强调空隙,与周围的公共空间进行整合,这一点在人行道的连续性、向外部(商店、咖啡馆、公共厕所)分配大型活动以及没有障碍物中得到了明确。这不仅是为了扩大建筑内部的城市公共生活,相反,也促进了艺术和文化传播到外部,使它们更加公开,也更加易于参与。

博物馆的每个封闭功能空间都与开放空间相辅相成,譬如阳台、广场或露台。这些空间除了作为公共空间或延伸的露天展览空间之外,也让使用者能在此短暂停留,或者欣赏城市风光。通过以下方式可以实现多种可能性:在剧院内做演讲、放映露天电影,或将广场当成舞台举办一场音乐会,以及让观众沿着剧院看台观看演出。

The Modern Art Museum of Medellin (MAMM) was founded in 1978 by a group of artists with the aim to arouse public interest on modern and contemporary art.
In 2006, Medellin established the reconversion of industrial land from SIMESA steel plant into a mixed used neighborhood called "Ciudad del Rio" (River City). The urban plan considered the dismantling of all factories to allow for generic high-rise blocks in a rigid scheme, leaving only the oldest nave -"Talleres de Robledo"- as the sole testimony of the industrial past. It was successfully refurbished by the mythical Grupo Utopia as the new venue of the Modern Art Museum on the 30th anniversary of its foundation.

In 2009, MAMM organized an international design competition for its expansion to house the requirements for a world class museum. The jury chaired by Spanish architect Federico Soriano and the public vote of those who visited the exhibited proposals awarded the first prize to the team of 51-1 from Peru and Ctrl G from Colombia.
The winning proposal was initiated from the application of Medellin's own informal settlement patterns of progressive growth at its hillsides. Piled on top of each other, the construction in the barrios steps up creating thousands of public interstices, where people exercise their urbanity in the most ingenious ways.

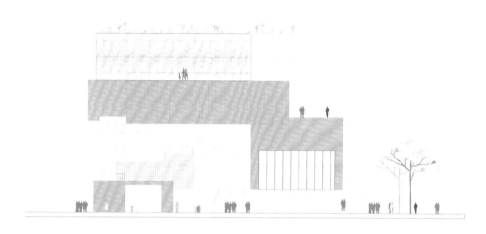

东立面 east elevation

北立面 north elevation

项目名称：Medellin Modern Art Museum Expansion / 地点：Medellin, Antioquia, Colombia / 建筑师：51-1 + Ctrl G / 主持建筑师：César Becerra, Manuel de Rivero, Fernando Puente Arnao - 51-1; Catalina Patiño, Viviana Peña - Ctrl G / 总协调：Isabel Dapena / 设计师：Sebastián Monsalve, Jorge Gómez, Eduardo Peláez
设计合作方：Luisa Amaya, Oscar Cano, Lucia Largo, María Camila Giraldo, Juan Camilo Arboleda, Nicolás Martínez, Favio Chumpitaz, Bruce Wong, Felipe Vanegas, Carolina Vélez, Luisa Echeverry, Juliana Vélez, Felipe Walter, Mónica Suárez, Sebastián Mejía, Camilo Martínez, Paula Mesa, Juan David Vargas, Luisa Lara, Juan José Riva, Juan Pablo Giraldo
施工阶段：Lina Durango, Laura Burgos / 结构工程师：Nicolas Parra G - CNI Ingenieros Consultores / 施工合作方：Ivan Dario Acevedo, Lorena Cañon, Carol Pavon, Santiago Parra / 总承包商：Conconcreto / 监理：Luis Guillermo Restrepo + cia s.a.s. / 客户：Museo de Arte Moderno de Medellín + Alcaldía de Medellin / 用地面积：2,982m² / 建筑面积：7,500m² / 总楼面积：1,176m² / 结构与材料：concrete - Mainframe, Columns, Beams and Walls; prefabricated concrete panels with different textures and slots - Exterior enclosures; metal structure and perforated metal plates as exterior enclosure - Last floor; steel - Exterior staircases and bent columns / 造价：US$ 9,210,000 without museum and theater tecnical dotations (lighting, cameras, etc) / 设计时间：2009.12—2011.11 / 施工时间：2013.8—2015.8
摄影师：©Federico Cairoli (courtesy of the architect)

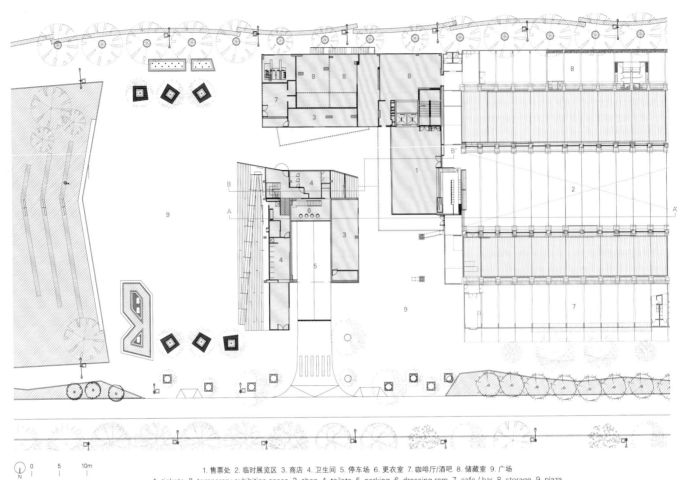

1. 售票处 2. 临时展览区 3. 商店 4. 卫生间 5. 停车场 6. 更衣室 7. 咖啡厅/酒吧 8. 储藏室 9. 广场
1. tickets 2. temporary exhibition space 3. shop 4. toilets 5. parking 6. dressing rom 7. cafe / bar 8. storage 9. plaza
一层 first floor

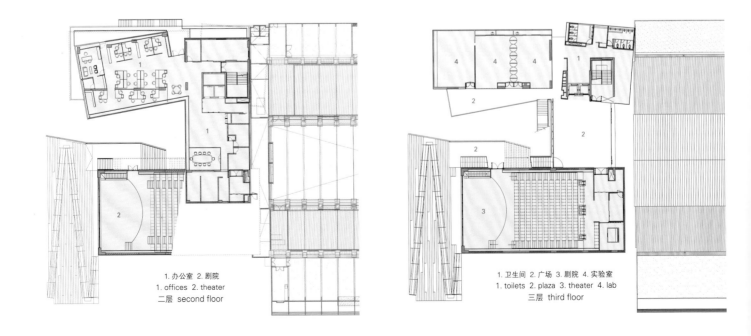

1. 办公室 2. 剧院
1. offices 2. theater
二层 second floor

1. 卫生间 2. 广场 3. 剧院 4. 实验室
1. toilets 2. plaza 3. theater 4. lab
三层 third floor

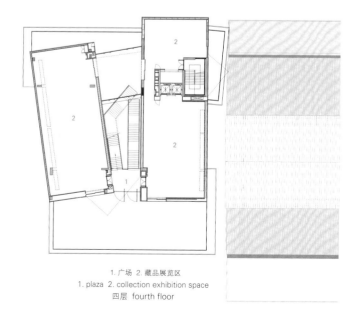

1. 广场 2. 藏品展览区
1. plaza 2. collection exhibition space
四层 fourth floor

1. 卫生间 2. 广场
1. toilets 2. plaza
五层 fifth floor

1. 临时展览区
2. 商店
3. 卫生间
4. 停车场
5. 更衣室
6. 咖啡厅/酒吧
7. 广场
8. 剧院
9. 藏品展览区

1. temporary exhibition space
2. shop
3. toilets
4. parking
5. dressing rom
6. cafe / bar
7. plaza
8. theater
9. collection exhibition space

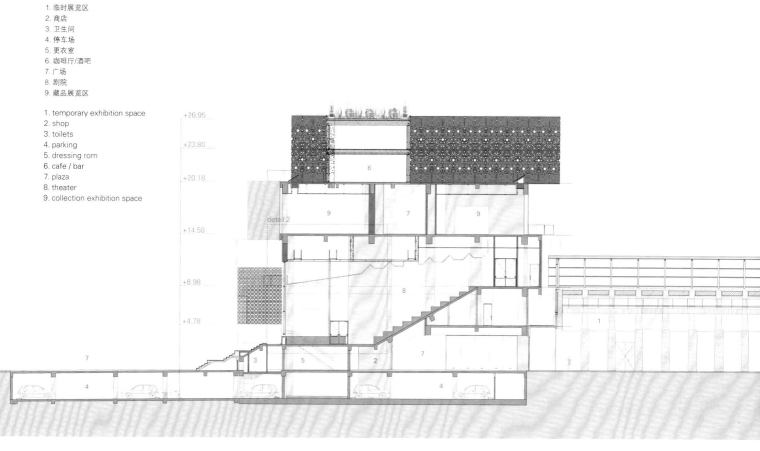

A-A'剖面图 section A-A'

1. 售票处
2. 临时展览区
3. 办公室
4. 广场
5. 藏品展览区

1. tickets
2. temporary exhibition space
3. offices
4. plaza
5. collection exhibition space

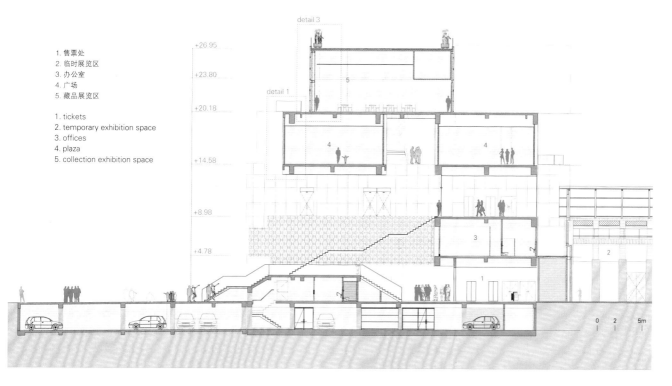

B-B'剖面图 section B-B'

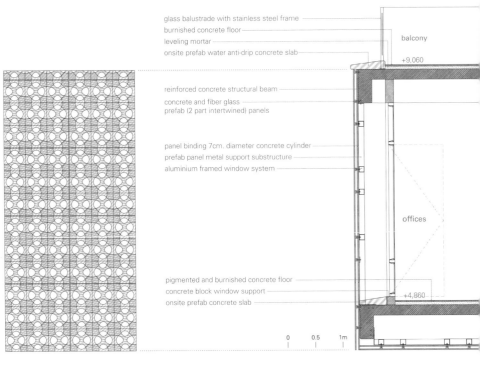

详图1 detail 1

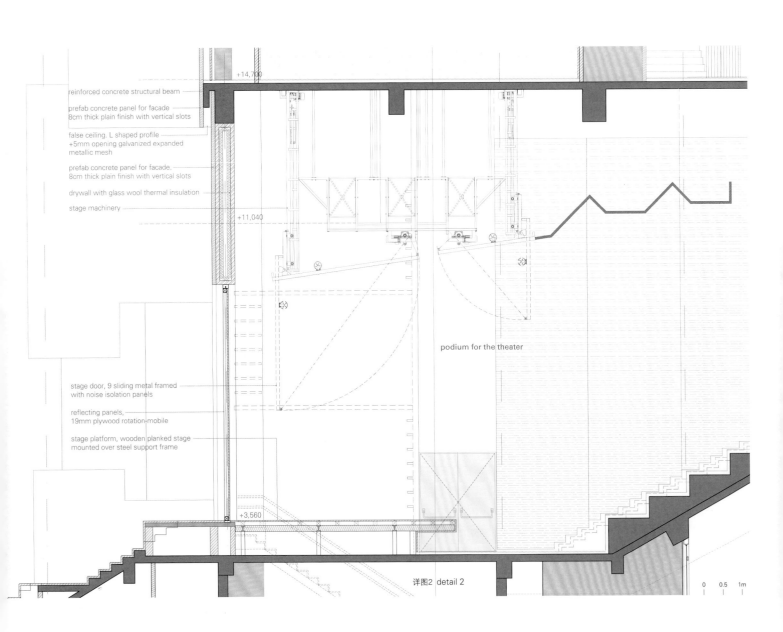

详图2 detail 2

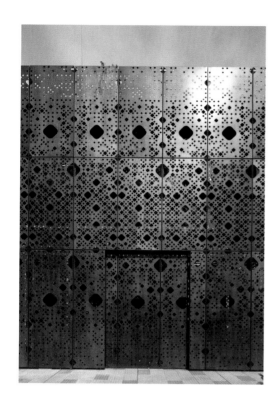

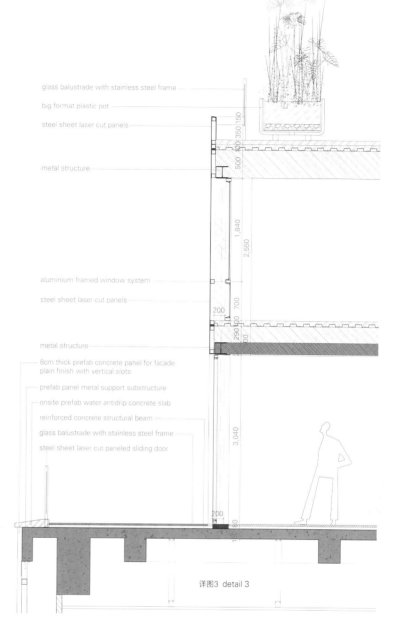

详图3 detail 3

A Village, Not a Building

Instead of a building, a vertical village was created. This way, as in the informal barrios of stacked houses in the hillsides, where one's roof is his / her upstairs neighbor's terrace, in the MAMM Expansion, the required programs are piled up making a cascade of public terraces connected by stairs. Inside is a promenade of art spaces. Outside, a vertical extension of the linear park of Ciudad del Rio is going up to the fifth floor: each one of these terraces can be seen as spatial reserves for the future growth of museum spaces.

Eleven Boxes and a Void

The museum is built in 5 levels. Besides new art galleries, it hosts laboratories, storage, offices, shops, cafés and a theater. These areas need to be closed and hermetic, with special light and geometric conditions. They are like boxes. Each one is closed by prefabricated concrete panels which are pierced, cut, and carved to reveal different colos, light and textures. Eleven boxes – of different sizes and heights – rotate and pile on top of each other, in a careful and strategic placement. This in-between empty space is a place where programs are mixed, and relations multiplied between

people and dialogues are allowed with the landscape.

Integration

The extension of the central nave into the center and inside the 51-1 + Ctrl G addition configures the new main entrance of the museum: a great void that connects old and new. It is intended an emphatic aperture and integration towards the surrounding public space, which is made clear in the continuity of the pavement, the allocation of great programmatic activity towards the exterior (shops, cafés, public toilets) and the absence of barriers. This is not only to extend the city's public life inside the building, but also to – inversely – make possible that art and culture propagates to the exterior, making them more public and accessible. Each closed program from the museum complements with an open space, being a balcony, a plaza or a terrace. These spaces, beyond configuring as public spaces or open-air extensions of exhibitions, allow for users to pause and gaze the city. Many possibilities are enabled with this: a lecture inside the theater, the projection of an open-air movie or the staging of a concert having the plaza as a stage and the audience along the theater tribune.

塞纳河音乐厅
La Seine Music Hall

Shigeru Ban Architects Europe

塞纳河音乐厅是一个多功能的音乐设施,位于巴黎西郊的塞金岛。这座岛曾因欧洲最大的汽车制造商——法国雷诺公司的旧制造厂而闻名。1992年工厂关闭后,原计划在此建一座由安藤忠雄设计的皮诺基金会博物馆,然而在开工前被皮诺本人取消了。后来,在私人融资计划(PFI)的支持下,该公司决定在岛上建造一座多功能的音乐设施,而坂茂建筑事务所欧洲办公室赢得了于2013年举办的设计竞赛。

该场地长330m,位于塞金岛逐渐收缩变窄的一侧。场地的长度和宽度几乎和艾菲尔铁塔一样。在这里,建筑师设计了一个总建筑面积为36500m²的多功能音乐设施(包括4000个座位的多功能音乐厅"大塞纳河"、1150个座位的古典音乐厅"大礼堂"、音乐学校、商店等)。客户要求为巴黎的西大门设计一些有"纪念性"和"象征性"的结构。建筑师没有创造出不同寻常的形状,而是创造了一颗珍贵的"宝石",即大礼堂。

大礼堂是葡萄园风格的古典音乐厅（座位环绕舞台），一共可容纳1150个座位，它的形状如同鸡蛋，表面覆盖着马赛克瓷砖，随着光线和视角的改变，这些瓷砖的颜色会由翠绿变为古铜色。这种马赛克瓷砖是为本项目特别制作的，灵感来自日本甲虫"TAMAMUSHI"。礼堂周围的门厅覆盖着木质结构的玻璃幕墙。设计竞赛要求使用3000m²的太阳能电池板。一张三角形的"帆"由太阳能电池板设计而成，而不是像通常那样将太阳能电池板放置在屋顶上。这个帆形结构被放置在礼堂的周围，它会根据太阳的光线移动，就像普通的帆随风移动一样。由于帆形结构处于运动状态，一直面向太阳，因此实现了高效的电力生产，也为礼堂的门厅创造了一处阴凉之所。礼堂内部的墙壁上则覆盖着波纹状的木条。

　　墙面木板条的排布方式考虑到了声学反射和吸收效果，用一模一样的木板条排列而成。各种各样的图案确保了声波的均匀扩散，让人们在所有的座位都能感受到同样的声学质量，也创造了不同的阴影。按照让·努维尔对整座岛的总体规划，这栋建筑被水边的裸露混凝土墙所包围，因为总体规划需要一些体现"野性"的东西，比如雷诺的前制造厂。与其外观形成对比的是塞纳河音乐厅的"宝石"，那正是大礼堂所在地。

　　在整座建筑中，商业轴得到了扩展，这也是在整座岛屿的总体规划中规定的。在不举办音乐会的时候，玻璃百叶窗（竖向闸门）和双折叠门是打开的，这样公众就可以在室内街道穿行了。进入建筑之后，游客可以沿着商店、售票处、餐馆、门厅和窗户散步，从窗户里还可以看到里面的排练室。这条街道引领游客来到"罗丹广场"，这是一个雕塑露台，位于岛的一端。

　　绿化带属于在总体规划中规定的另一条轴线，一直延伸到塞纳河音乐厅。游客可以爬上壮观的外部楼梯，进入位于多功能音乐厅"大塞

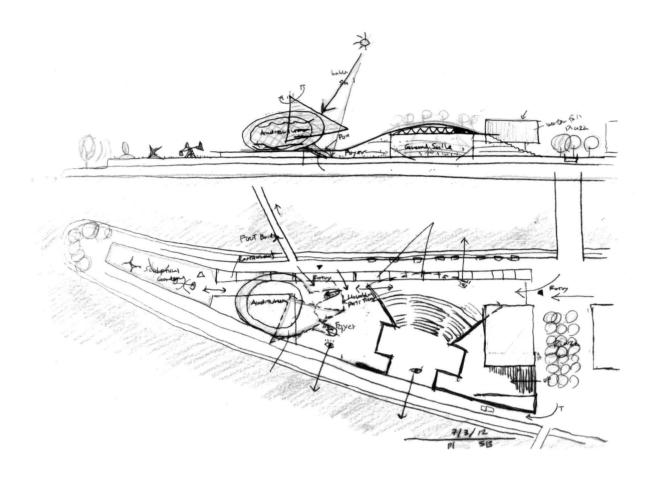

项目名称：La Seine Music Hall / 地点：Seguin Island in Boulogne-Billancourt (Suburb of Paris), France / 建筑师：Shigeru Ban Architects Europe + Jean de Gastines Architectes / 项目团队：Shigeru Ban, Philippe Monteil, Nicolas Grosmond, Geoffroy Boucher, Alessandro Boldrini, Masashi Maruyama, Sara Lazzarin, Amélie Fritzlar, Alexis de Dumast, Veronica Arianna, Mathieu Chapus, Patrick Allan, Marc Ferrand, Grégoire Defrance, Kazuhiro Asami, Claude Hartmann - SBAE; Jean de Gastines - JDGA
客户的项目经理：LAURENT Jean-François / 设计团队的项目经理：Grégoire Defrance - competition phase; Nicolas Grosmond - development phase; Philippe Monteil - construction phase / 总承包商：VIEILLARD Thibaut - Bouygues Bâtiment Ile-de-France
客户：Conseil Général des Hauts-de-Seine / 顾问：SETEC TPI - General; Blumer-Lehmann AG - Timber; ARTELIA – MEP; dUCKS scéno - Stage and hall equipment design; LAMOUREUX ACOUSTICS – Acoustic; NAGATA ACOUSTICS - Hall Acoustic; RFR - Façade until CD; T/E/S/S atelier d'ingénierie – Façade from CA; Bassinet Turquin Paysage – Landscape / 预计施工造价：US$ 243 M
建筑师设计费：US$ 12M / 用地面积：23,000m² / 建筑面积：16,000m² / 总建筑面积：36,500m² / 施工时间：2013.3—2016.10
竣工时间：2016.12 / 摄影师：©Didier Boy de la Tour (courtesy of the architect)

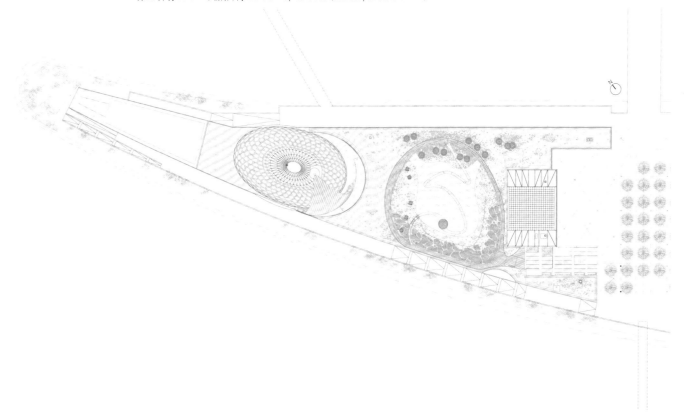

纳河"上方的屋顶花园。

这并不是客户提出的要求,而是建筑师想要建造像博物馆和音乐厅这种对所有人都开放的公共设施,换句话说,建筑师想吸引不习惯参观此类设施的市民前来,当然也欢迎音乐和艺术爱好者。塞纳河音乐厅的室内街道和屋顶花园就是实例。立面上还有一张巨大的LED屏幕,面向前院。建筑师希望,这张欧洲最大的屏幕能让没有票的人也可以欣赏到在两个大厅内上演的戏剧。

La Seine Music Hall is a multi-purpose musical facility, located on Seguin Island in the western suburbs of Paris. This island was known for the old manufacturing plant of Renault, the largest European automobile manufacturer. After the factory was closed in 1992, it was planned to build a museum of the Foundation Pinault designed by Tadao Ando. However, the plan was canceled by Mr. Pinault himself right before the construction started. Later, it was determined to build a multi-purpose musical facility on this island under the Private Finance Initiative (PFI) and Shigeru Ban Architects Europe won the design competition held in 2013.

The site is 330m long, at a narrowing edge of the Seguin Island. The Eiffel Tour happens to be almost as long and as wide as the site. Here, a multi-purpose musical facility was designed with the total floor area of 36,500m² (including a multi-purpose concert hall "Grande Seine" with 4,000 seats, a classic concert hall "Auditorium" with 1,150 seats, music school, shops etc,). Designing something "monumental" and "symbolic" for the west gate of Paris was required by the client. Instead of making an unusual shape building, the architect rather created a precious "jewel", the auditorium. The auditorium is a vineyard style classic concert hall (the seating surrounds the stage) which holds 1,150 seats in total. Its form is egg-like, and its surface is covered with

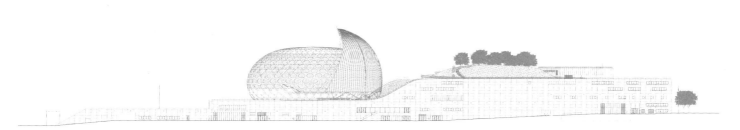

西南立面
south-west elevation

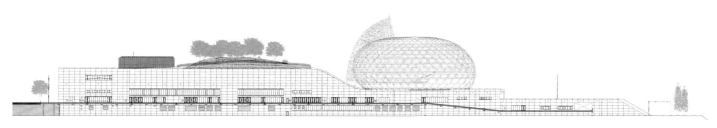

东北立面
north-east elevation

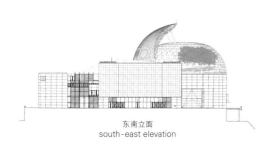

东南立面
south-east elevation

西北立面
north-west elevation

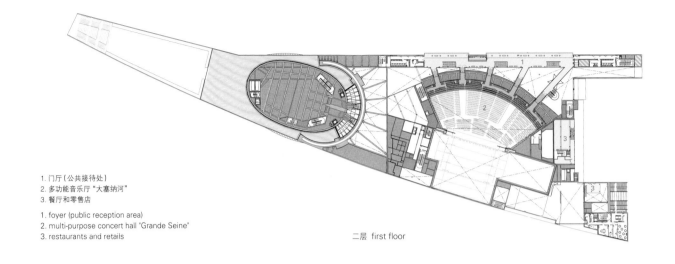

1. 门厅（公共接待处）
2. 多功能音乐厅"大塞纳河"
3. 餐厅和零售店

1. foyer (public reception area)
2. multi-purpose concert hall "Grande Seine"
3. restaurants and retails

二层 first floor

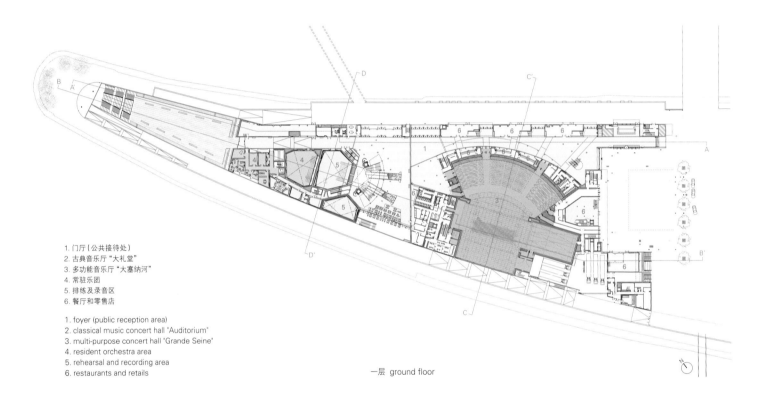

1. 门厅（公共接待处）
2. 古典音乐厅"大礼堂"
3. 多功能音乐厅"大塞纳河"
4. 常驻乐团
5. 排练及录音区
6. 餐厅和零售店

1. foyer (public reception area)
2. classical music concert hall "Auditorium"
3. multi-purpose concert hall "Grande Seine"
4. resident orchestra area
5. rehearsal and recording area
6. restaurants and retails

一层 ground floor

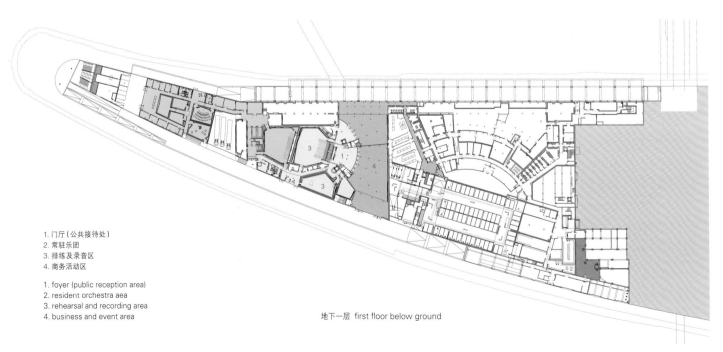

1. 门厅（公共接待处）
2. 常驻乐团
3. 排练及录音区
4. 商务活动区

1. foyer (public reception area)
2. resident orchestra aea
3. rehearsal and recording area
4. business and event area

地下一层 first floor below ground

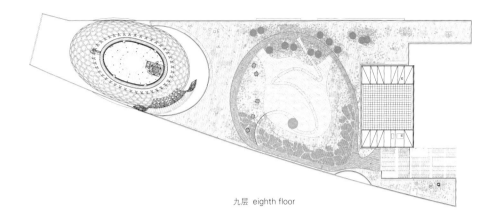

九层 eighth floor

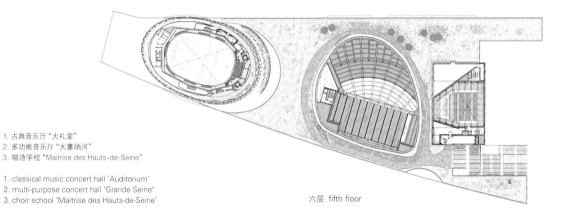

1. 古典音乐厅 "大礼堂"
2. 多功能音乐厅 "大塞纳河"
3. 唱诗学校 "Maitrise des Hauts-de-Seine"

1. classical music concert hall "Auditorium"
2. multi-purpose concert hall "Grande Seine"
3. choir school "Maitrise des Hauts-de-Seine"

六层 fifth floor

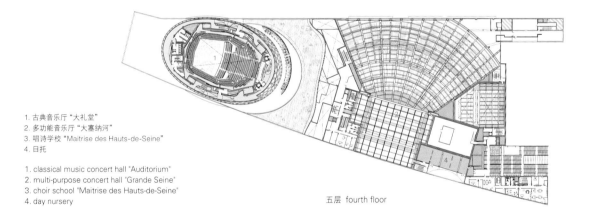

1. 古典音乐厅 "大礼堂"
2. 多功能音乐厅 "大塞纳河"
3. 唱诗学校 "Maitrise des Hauts-de-Seine"
4. 日托

1. classical music concert hall "Auditorium"
2. multi-purpose concert hall "Grande Seine"
3. choir school "Maitrise des Hauts-de-Seine"
4. day nursery

五层 fourth floor

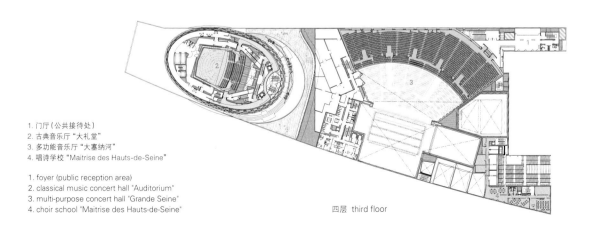

1. 门厅（公共接待处）
2. 古典音乐厅 "大礼堂"
3. 多功能音乐厅 "大塞纳河"
4. 唱诗学校 "Maitrise des Hauts-de-Seine"

1. foyer (public reception area)
2. classical music concert hall "Auditorium"
3. multi-purpose concert hall "Grande Seine"
4. choir school "Maitrise des Hauts-de-Seine"

四层 third floor

mosaic tiles whose color changes from emerald green to bronze-red according to the lighting and the angle of vision. This mosaic tile was especially produced for this project, inspired by the Japanese beetle called "TAMAMUSHI". The foyer around the auditorium is covered with glass claddings with wooden structure. The use of 3,000m² solar panel was required for the competition. Instead of ordinarily placing the panels on the top of the roof, a triangle "sail" was designed by using solar panels. This sail is placed around the auditorium, and it moves according to the sun lights, just as an ordinary sail moves according to winds. Thanks to its movement, the sail is facing the sun all the time, which allows an efficient power production and also creating a shade for the foyer of the auditorium. Inside the auditorium, the wall was covered with corrugated wooden stripes. Several typographies have been conceived by composing same single module differently concerning acoustic reflection and absorption. The variety of patterns ensures a homogeneous diffusion of the sound wave and an acoustic quality at all the seats and also creates different shadows. Following the Jean Nouvel's master plan of the entire island, the building is surrounded by undressed/exposed concrete wall along the water's edge, since the master plan required something "brutal" like the former manufacturing plant of Renault. Contrast to its exterior appearance, a "jewel" of La Seine Musicale, is where the auditorium was created.

Throughout the building the commercial axis has been extended which was defined in the master plan of the entire island. When there is no concert, Glass Shutter (Porte Guillotine) and Bi-folding Door (Porte Pliante) open up so that the public can go across the Interior Street (Rue Intérieure). Once inside, the visitors can walk along shops, ticket office, restaurants, foyer, and windows with which you can see inside the rehearsal rooms. This street leads visitors to the 'Place Rodin' a sculptural terrace placing at the point of the island.

The green belt, the other axis defined inside the master plan extends until La Seine Musicale. The visitors can go up the monumental exterior stairs to access to the rooftop garden placed above Grande Seine.

It is not a client's request but the architect's desire to build public facilities such as museums and concert halls which are open for everyone; in other words, the architect wanted to invite citizens who are not used to visiting such facilities, along with music and art devotees. The Interior Street and the rooftop garden of La Seine Musicale are some examples. There also is a Giant LED screen on the facade facing Parvis. The biggest screen of Europe was placed with the architect's hope that people who don't have tickets also can enjoy the plays taking place inside the two halls.

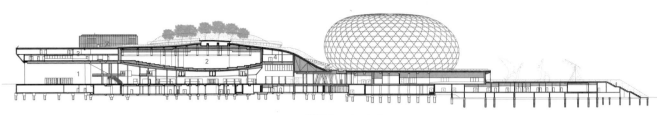

A-A'剖面图 section A-A'

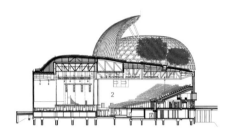

C-C'剖面图 section C-C'

1. 门厅（公共接待处）
2. 多功能音乐厅 "大塞纳河"
3. 唱诗学校 "Maitrise des Hauts-de-Seine"
4. 商务活动区
5. 餐厅和零售店

1. foyer (public reception area)
2. multi-purpose concert hall "Grande Seine"
3. choir school "Maitris des Hauts-des-Seine"
4. business and event area
5. restaurants and retails

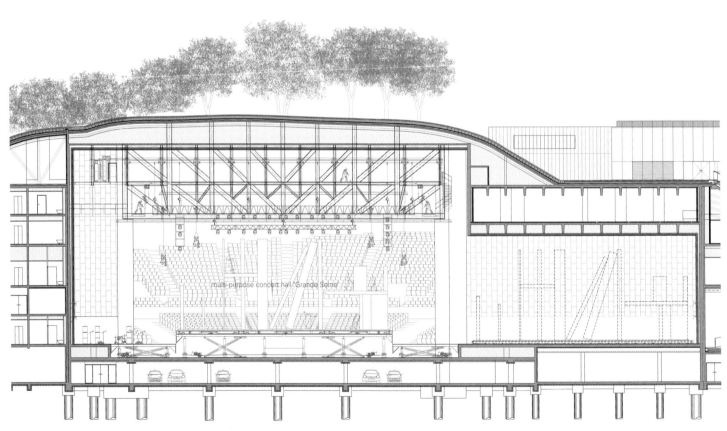

详图1 detail 1

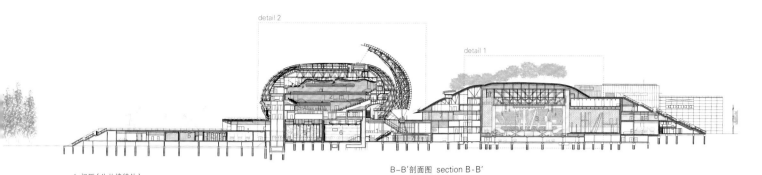

B-B'剖面图 section B-B'

1. 门厅（公共接待处）
2. 古典音乐厅"大礼堂"
3. 多功能音乐厅"大塞纳河"
4. 常驻乐团
5. 排练及录音区
6. 餐厅和零售店
7. 商务活动区

1. foyer (public reception area)
2. classical music concert hall "Auditorium"
3. multi-purpose concert hall "Grande Seine"
4. resident orchestra aea
5. rehearsal and recording area
6. restaurants and retails
7. business and event area

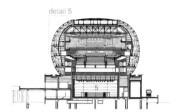

D-D'剖面图 section D-D'

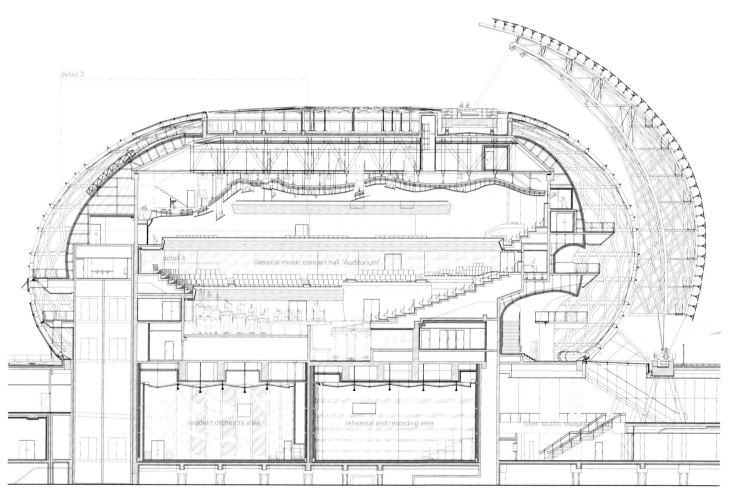

详图2 detail 2

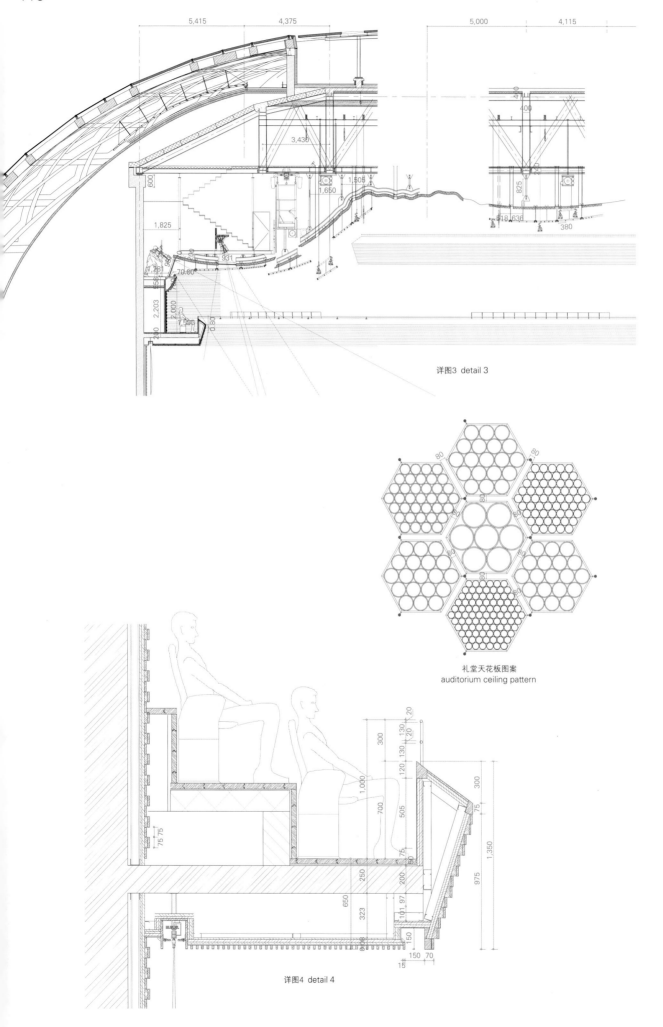

详图3 detail 3

礼堂天花板图案
auditorium ceiling pattern

详图4 detail 4

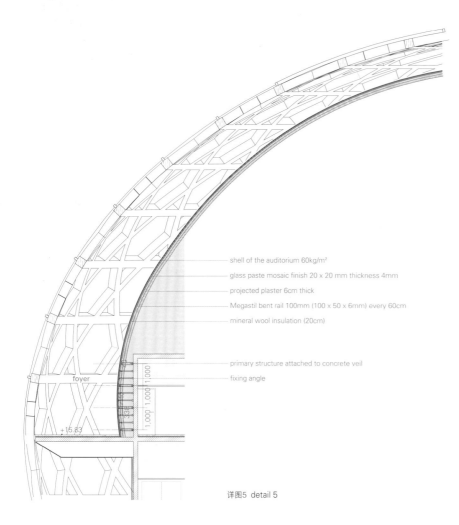

- shell of the auditorium 60kg/m²
- glass paste mosaic finish 20 x 20 mm thickness 4mm
- projected plaster 6cm thick
- Megastil bent rail 100mm (100 x 50 x 6mm) every 60cm
- mineral wool insulation (20cm)
- primary structure attached to concrete veil
- fixing angle

详图5 detail 5

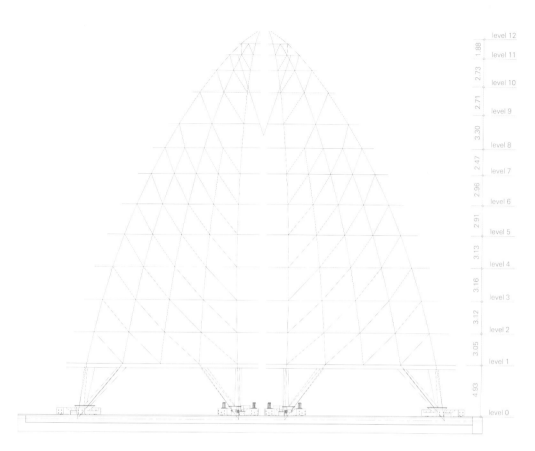

太阳能板立面
solar panel elevation

学习中的城市

把学校当成社交游乐场
The School as Social Playground

中小学是孩子们从幼儿园小宝宝成长为小学生、中学生的地方。教育的第一个阶段需要特别注意，尤其是在建筑空间上：空间的形式、氛围、材料、颜色及其功能、启发和招待能力。建筑师要面对孩子的需要，学校的空间要满足他们的特殊需求。

在幼儿园的房间里，我们可以经常发现与幻想世界的联系，而在学校的建筑里，空间是按照学习计划的要求来设计的，主要发生在教室和体育活动的空间里。然后可能有食堂和自助餐厅、睡觉和休息的空间、室外或室内的运动场地。

除了这些空间之外，本书所选中的学校建筑有一个共同的重要特征，那就是"关系空间"或"公共

Primary and secondary schools are the places where the children experience their growth from kindergarten kids onto little students. This first stage in education requires some special attention, and especially so in the building spaces: their shape, atmosphere, materials, colors, as well as their functions, their ability to inspire and to invite. The architect confronts with the necessities of the kids, and the school spaces need to answer their specific demands.

While in kindergartens' rooms we could frequently find links with the fantasy world, in school buildings the spaces are modeled along the requirements of learning programs, that mainly take place in classrooms and physical activity spaces. Then there may be canteen and cafeteria rooms, spaces for sleep and rest, outer or inner sport fields.

哥本哈根诺德海文国际学校_Copenhagen International School Nordhavn/C.F. Møller Architects
丹麦Frederiksbjerg学校_Frederiksbjerg School/Henning Larsen Architects + GPP Arkitekter
斯蒂芬·佩尔斯基金会高中_Stephen Perse Foundation Senior School/Chadwickdryerclarke Studio
斯科夫巴肯学校_Skovbakken School/CEBRA
马德里德语学校_German School Madrid/Grüntuch Ernst Architects
学习中的城市——充满社交生活的学校_Learning Cities – Schools of Social Life/Andrea Giannotti

空间"的存在。它们没有明确的功能,只是把不同的房间连接在一起,从而为学生提供一个见面、交谈和玩耍的地方,虽在学校内部,但在教室之外。

公共空间在学校建筑中是比较新的存在,在学生们准备面对社会生活的过程中扮演着主要角色,并在与他人的关系中塑造了他们的个性。这一特点提出了学校的关联空间及其文化和地理环境之间的联系,以强调项目在功能、材料和课程可持续性等方面的品质。这些方面都对学生对个人以及社会生活的理解产生了强烈的影响。

Besides these spaces, the school buildings of this selection have one important feature in common: the presence of the "relation spaces", or "common spaces". They have no definite function but connecting the various rooms together, and thus offering the students a place to meet, talk and play, inside the school, but outside the classrooms.
The common space, relatively new in school buildings, has a central role in preparing the students to confront social life, and in shaping their individuality in relation with the others. This feature proposes a link between the school's relation spaces and their cultural and geographical context, as to highlight the qualities of the projects in terms of functions, materials, and of course sustainability. All of these aspects have strong influence on the students' personal and social understanding of life.

学习中的城市——充满社交生活的学校
Learning Cities – Schools of Social Life

Andrea Giannotti

要从建筑和城市两个层面对中小学进行审视，首先要探究的是，对于孩子来说，学校究竟是什么？有些孩子会回答学校是学习、实验和探索的地方；另一些孩子可能会说它是玩耍和娱乐的地方；还有一些孩子会说它是交朋友的地方。事实上，一所供孩子们使用的学校可能包含以上所有含义。

在设计学校的空间时，建筑师必须考虑到所有这些情况，因为学校是孩子开始在家庭之外与他人建立联系并构建自己个性的地方。因此，这种空间的设计者应该非常关注孩子，孩子才是学校最主要的"使用者"：他们还没有完全长成，会吸收来自所居住空间的各种迹象。本书所选用的学校项目正是这种设计研究的明确成果。从建筑层面来说，幼儿园的房间往往与幻想世界有联系，而在学校建筑中，空间是根据学习计划形成的，主要在教室和专门的活动空间内进行。根据具体的项目功能要求，在小学和中学可能会有食堂或自助餐厅、休息或睡觉的房间、室外或室内运动场地。这些学校实际上提供了一套符合少年儿童需要的功能空间。

将比喻延伸开去，我们可以想象一下，学校为学习、吃饭、活动和玩耍、休息和睡觉提供了空间，就像城市——或者任何有组织的环境，就像乡下的一个村庄——为一个群体提供了空间。

因此，"学习中的城市"这个标题是指学校空间与城市空间的对应关系。所有这些学校项目都具有与一般城市相同的各种空间：学校的特定学习场所是教室，而城市的特定学习场所是学校、大学、博物馆和图书馆；将学习功能扩展到生产活动当中，在学校我们有工作坊或实验室，而在城市有办公室、手工艺商店和工厂；至于饮食方面，在学校我们可能会找到一个食堂或自助

To start a review about primary and secondary schools on architectural and urban level, we should first inquiry what exactly is the school for a child. Some children could answer that school is the place of learning, experimenting and exploring; others could mention it as the place of playing, entertainment and fun; some others could say it is the place for making friends. In fact, a school for a kid might have all of these meanings all together.

While designing the spaces of a school, the architect must take in consideration all these interpretations, because the school is the place where the child starts building up his own personality, in relation with others and outside the family. Therefore, the designer of such spaces should pay great attention to the kid, the school's uttermost "user": the child is not yet completely formed, and he or she absorbs the suggestions that come from the spaces he lives in. The schools in this selection are the explicit result of such design research. On architectural level, while kindergartens' rooms often bear links with the fantasy world, in school buildings the spaces are shaped according to learning programs, which mainly take place in classrooms and dedicated activity spaces. Depending by the particular program, in primary and secondary schools there may be a canteen or cafeteria, rooms for rest or sleep, outer or inner sport fields. These schools provide in fact a set of functional spaces corresponding to the necessities of the child-student.

Stretching a metaphor, we could imagine that the school provides spaces for learning, eating, moving and playing, rest and sleeping, in the same way as a city – or any structured environment, as a village in the countryside – does for a community of people.

The title "Learning Cities" refers thus to the correspondence between the school spaces and the city spaces. All these school projects are equipped with the same variety of spaces of a generic city: specific learning places in a school are the classrooms, while in a city they are the schools, universities, museums and librar-

餐厅，那在城市里就变成了餐厅和酒吧；对于体育锻炼来说，学校里有体育馆和室外运动场，在城市里也有，占据了更大的公共区域，比如公园和运动场；说到娱乐，仍然与学习有关，在学校有特定的房间作为礼堂和剧院，而在一个城市中，娱乐场所的名单就长了，包括剧院、电影院、音乐厅等。最后但同样重要的是，在学生居住的学校，我们看到了宿舍或用于睡眠的私人房间，在城市里当然是大片的独栋住宅和住宅楼了。

既然有了这个比喻，"学习中的城市"这个标题在这里就可以解释为，学校被设想为"学习型城市"，学校给学生提供了他们当前生活所需的各种空间，并延伸到他们作为城市居民的未来生活。

然而，如果一个城市不包含任何社交空间，那么这座城市中的生活是不完整的。

家庭住宅之外的人际关系是人类一种特定的基本需求，事实上，大多数人都生活在社区里，无论大小，这些社区都形成了核心群体，旨在实际打造居住空间，塑造环境，形成了村庄、城镇，乃至城市。

任何一个村庄或城市都有许多不同的社会或关系空间：它们共同构成了公共空间，正好与住宅的私人空间相对应。在城市中，自一座城市有历史开始，社会空间就是街道和广场，有时则是特殊的公共建筑，它提供了一个屋檐，每个人都可以在屋檐下遇见任何人。

但公共空间的独特性，例如，街道或广场，一直以来都是一种非专用的功能。虽然我们承认街道是专门用于人和车辆的流通的，但这并不是其唯一的功能：在街道和广场上，人们往往只是见面、聚会、聊天、玩耍、闲逛、吃吃喝喝。在公共空间也有相

ies; extending the learning program to production activities, in schools we have workshops or laboratories, while in a city there are offices, handicraft shops and factories; as for eating, at school we may find a canteen or cafeteria, that in the city become restaurants and bars; for sport practice, in schools there are gyms and outdoor playgrounds, which are also in the city, along with bigger public areas like parks and sport fields; for entertainment, still associated with learning, in schools there are specific rooms as auditorium and theaters, in a city the list is so long as to include theaters, cinemas, concert halls and so on. Last but not least, in the schools where students live in we see dormitories or private rooms for sleeping, that in the city are of course the large carpet of houses and housing complex.

Granted this metaphor, the title "Learning cities" can be here interpreted in the sense that schools are conceived to be "cities of learning", where the students are given all the kinds of spaces needed for their current life, and by extension for their future life as inhabitants of a city.

Still, life in a city is not complete if it does not include any space for social interaction.

Relations set outside the family house are a specific and fundamental need of humans, who in fact live mostly assembled in communities, large or small, that form then the core group meant to physically build the living space, shaping the environment: villages, towns, cities.

In any village or city there are many different social or relational spaces: together they form the public space, opposed to the private space of the house. In the city, since the beginning of urban history, the social spaces are the streets and the squares, and sometimes special public buildings to provide a roof under which everyone can meet anyone.

But the distinctive character of the public space, as the street or the square, has always been the one of a not-exclusively dedicated function. Even though we admit that streets are dedicated to the circulation of

沃尔特·格罗皮乌斯设计的德绍的包豪斯建筑，德国，1925年
Bauhaus in Dessau by Walter Gropius, Germany, 1925

对的表达自由。唯一的限制是要遵守塑造每个社区公共生活的法律和法规。因此，公共空间是居民进行社会交往最多的场所。

假设孩子们是学校城市的公民，这就支持了学校作为城市的比喻，确定学校建筑内公共空间的位置是很重要的。它是所有学生、教师和工作人员共同使用的空间，除了连通其他所有房间之外，它不应该设有专门的功能。

很明显，我们可以在本书所选取的这些项目中找到公共空间。除了空间、灯光和材料的特点外，人们对这些空间的使用也能让我们认出它们：在照片中，我们看到学生们见面聊天，孩子们玩耍、看视频，一些人在读书，一些人凑在一起吃饭。在这些学校中，我们都注意到公共休息室的功能是向所有人开放的，几乎可以派上所有用场：丹麦的三个项目有高大的日光浴中庭，有供孩子和老师坐着聊天的宽阔楼梯，或者有可供玩耍的体育锻炼设施；西班牙的这所学校提供了户外公共空间，上方由屋顶覆盖。尽管所选用的材料不同——混凝土、金属板、木材和砖块，但是它们都使用了一种建筑语言，邀请学生们享受这个开放空间作为他们课后的聚会场所。

简而言之，孩子们在公共空间里与他人建立关系，尽管他们的活动有老师在旁监看，但大多数情况下是不受干扰的，而且在教室和家庭之外，他们的身心都是如此自由自在。然而，这些公共空间并不总是在学校里。从20世纪开始，建筑师在开始设计学校建筑时，就不再将它设想成简单的教室集合了。

迅速回顾一下整个20世纪的主要学校建筑，即使是最先进的设计学校——由麦金托什设计的格拉斯哥艺术学院（1909

people and vehicles, this is not an exclusive function: on the streets and squares, often people just meet, gather, talk, play, hang out, drink and eat. There is a relative freedom of expression in the public space. The only constraint is to observe laws and regulations that shape the public life in every community. Thus, public spaces are the places that inhabitants use at most for their social interaction.

Assuming that children are the citizens of the school-city, that is endorsing the metaphor of the school as a city, it is important to locate the public space within the school building. It is the space of common use, shared between all students, teachers and staff, and it should not have a specifically dedicated function, aside connecting all the other rooms.

We can find the common spaces, well evidently, in this selection of projects. They are recognizable, besides their spatial, lighting and materials features, also by the use that people do of them: in the photos we see students meeting and talking, children playing, watching videos, some reading, some eating together. In each of these schools we notice that the function of the common room is open to every use, mostly unconstrained use: the three projects in Denmark have large, high-ceiling, sunbathed atriums, with wide stairs for kids and teachers to sit and talk, or physical training facilities to play with; the one in Spain offers the common space outdoor, covered by a roof, and they use an architectural language that, although with different materials – as concrete, metal panels, wood and bricks - invites the students to enjoy this open space as their after-class meeting place.

In short, children are establishing their relations with the others in the common spaces, mostly controlfree even though observed, yet physically and mentally outside both the classroom and the family house.

These common spaces, though, have not always been in the schools. It is a peculiarity of the past century to start conceiving the school building as more than a simple collection of classrooms.

朔尔兄妹中学，汉斯·夏隆，德国吕嫩，1962年
Geschwister-Scholl Secondary School by Hans Scharoun, Lünen, Germany, 1962

阿姆斯特丹市政孤儿院，阿尔多·凡·艾克，荷兰，1960年
Amsterdam Municipal Orphanage by Aldo Van Eyck, The Netherlands, 1960

年），或格罗皮乌斯在德绍设计的包豪斯建筑（1925年），也有臭名昭著的案例：除了教室和实验室，几乎没有考虑公共空间。如果说存在类似于当代公共空间的关系空间的话，这个空间也都被放在了建筑外部，在最好的情况下是在庭院或学校花园里。

从20世纪50年代末到60年代，随着一些学校和教学体制的改革，我们发现一些学校有了中心中庭或宽阔的公共区域。这种情况阐明了对沃尔特·格罗皮乌斯设计的学校项目的观察，格罗皮乌斯过去设计的包豪斯学校建筑仍旧具有革命性，但缺乏社交空间，后来改成了如今以公共中庭为中心、四周环绕教室的学校建筑布局，正如他在柏林设计的Gropiusstadt学校（1969年）那样。

我们发现，另外两个建成于20世纪60年代初的重要学校项目为中心空间功能的转变奠定了基础。汉斯·夏隆在1958年至1962年间设计并完成了吕嫩中学：在该项目的平面设计中，一个有机的连通区域的布局得到拓宽，成为宽大的中庭，而不是简单的走廊。在同一时期，从1957年到1960年，荷兰建筑师阿尔多·凡·艾克在阿姆斯特丹Kindertehuis孤儿院的设计中尝试了结构主义原则：其平面设计由模块化的单元——房间和庭院组成，但是也将连通空间扩大，变成适合停留、社交的公共区域。在这种情况下，孩子们注定要生活在学校里，因此在宿舍和教室之外提供交流空间的必要性就更强了。

在当前的学校里，我们承认公共空间的存在，但在前几代学校中却很少发现它们的身影。这是因为设计原则是顺应时代、文化变迁和社会发展的：在20世纪中后期，教师和教学不再是学校功能的重要角色，取而代之的是学生及其学习环境。

A quick look at the main school buildings throughout the XX century highlights that even the most advanced designed schools – the Glasgow School of Art (1909) by Mackintosh, or the Bauhaus in Dessau (1925) by Gropius, as notorious examples – took little concern of the common spaces besides the classrooms and laboratories. If there was a relation space similar to the contemporary common space, this was placed outside the building: in best cases in the courtyard or school garden.

Starting from the late 50s and 60s, with the reform of several schooling and teaching systems, we find some schools with spaces such as central atriums or wide common areas. It is explicative of this context the observation of the school projects by Walter Gropius, who passed from the absence of free-function relation spaces inside the still revolutionary Bauhaus (1925) to a school layout centered on the common central atrium with classrooms all around, as it is in the Gropiusstadt schools in Berlin (1969).

We found two other important school projects completed in the first years of the 60s that paved the ground for this shift in the function of central spaces.

Hans Scharoun designed and completed the Secondary School in Lünen between 1958 and 1962: in the plan, an organic layout, the connective and circulation areas got widened until becoming a large atrium rather than a simple hallway. In the same years, between 1957 and 1960, Dutch architect Aldo Van Eyck confronted the design of the Amsterdam orphanage Kindertehuis experimenting with the Structuralism principles: the plan is a combination of modular units, rooms and courtyards, but the connection spaces are also enlarged as to become common areas to stay in, and use for, social interaction. In this case, the children were meant to live in the school, therefore it was stronger the necessity to provide relational spaces besides the dormitories and the classrooms.

We acknowledge the presence of common spaces in the modern and contemporary schools, but rarely find

斯蒂芬·佩尔斯基金会高中,英国
Stephen Perse Foundation Senior School, UK

斯科夫巴肯学校,丹麦
Skovbakken School, Denmark

  这意味着,在当代学校中,尤其是在本书所选择的几所学校中,设计的焦点显然是孩子,鼓励他们用信息来满足自己的好奇心,塑造自己,并探索社会交往行为。这些学校提供了适合孩子们所有活动的空间,在这里,孩子们建立了自己的社会精神:他们受邀使用那些位于校园中央的、未指定功能的开放公共区域,将之当成互动游乐场,产生共同的经历。

  从这个意义上说,我们可以把本文标题"学习中的城市"理解为"学习在城市中生活",也就是学习在社区中生活。除了作为一个"学习型城市",如今的学校更应该有意识地把自身当成孩子们体验社会、集体生活的第一个地方。

  当代的学校在空间设计中必须遵守这一角色,我们可以称之为"社会责任";同时,它们必须考虑在城市当中的角色或者在景观中的位置;显然,它们也必须特别关注整体建筑的可持续性,包括节能和产能策略。

  因此,校园与外部环境和地理位置的关系在设计选择中起着决定性的作用。将可持续性发展原则与文化和地理环境相匹配,就产生了建筑的地方特色。尽管这些学校采用了与儿童专用空间相同的方式,更有利于孩子们的社会互动,但是建筑必须声明其特殊性,表明与其他建筑的不同之处。

  正如建筑师C.F. Møller所说的:"现代教育建筑旨在将学校建筑与城市环境中的公共空间联系起来。"因此,本书案例中的每一所学校都有其独特的方式与城市或自然景观产生联系:哥本哈根诺德海文国际学校(132页)位于一个港口码头,设计了临水的露台和大型中庭,而斯科夫巴肯学校(182页)精心设计了屋顶的线条,与居民区的建筑互相呼应;Frederiksbjerg学校(148

them in the previous generations of schools. That is because design principles follow the times, the cultural shifts and social developments: in the second half of 1900, the students and their learning replaced the professors and their teaching, as the prominent role of the school programs.

This means that in contemporary schools, and especially the ones in this selection, the main focus in design has clearly been the child, encouraged to fill up his curiosity with information, to build his person, and also to explore social interaction. These schools provide the space adapted to all the activities, and in here the children build up their social spirit: they are invited to use those central, open, un-specified common areas, to make them a sharing experience, as a social playground.

In this sense we could interpret the terms of the title: "Learning Cities" as "learning to live in a city", that is learning to live in a community. Besides being a "city of learning", the school nowadays should offer itself consciously as the first place where the kids experience their social, collective life.

Contemporary schools have to comply with this role, which we could call "social responsibility", in the spaces design; at the same time, they have to consider their urban role or their insertion in the landscape; and obviously, they have to take special concern of the overall building sustainability, both in energy saving and producing strategies.

Thus, the relation with the external context and geographical location plays a decisive role in leading the design choices. From matching the sustainability requirements with the cultural and geographical environment comes the local personality of the building. Even though these schools share the same approach as spaces dedicated to the children, favoring their social interaction, the building must declare its own specificity, its difference from the others.

As the architect C.F. Møller said, "the modern educational architecture is designed to link the school premises

哥本哈根诺德海文国际学校，丹麦
Copenhagen International School Nordhavn, Denmark

Frederiksbjerg学校，丹麦
Frederiksbjerg School, Denmark

页）则增加了其主要体量，创造了位于地下的城市公共空间；斯蒂芬·佩尔斯基金会高中（166页）设有隐秘的景观庭院和灵活的公共空间，用于突发活动和小组学习；马德里的德语学校（196页）则采用了混凝土进行设计，但这所学校通过开放或半封闭的空间和庭院，以热情友好的方式创造了属于他们的城市花园。

关于环境和可持续发展战略方面，值得注意的是，丹麦的学校是如何为所有的社会和体育活动设置大型内部中庭的，而出于气候原因，西班牙的学校则将这些房间设计成了开放的空间。

可持续性不仅是成年人和专业人士所担心的技术问题，而且向儿童说明这一问题也将帮助他们在生活中培养一种可持续发展的意识。可持续发展属于我们可以称之为"文化责任"的更大范畴的话题，包括学校建筑所能展示和提供给孩子们的一切，都是为了丰富他们的评价体系和知识。

从这个角度来解读，现代学校建筑设计的主要课题仍然是帮助孩子在未来的生活中树立社会和文化责任感。

学校和城市之间的比喻只是设计问题复杂性的一个例证。然而早在1960年，阿尔多·凡·艾克在阐述他设计的Kindertehuis孤儿院的理念时，就使用了这句话："一个小世界里有一个大世界，在城市的小房子里有一个大世界，城市就是一座房子，一座为孩子而建的房子。"着重强调这句话，也就是在强调本书所选学校案例的设计概念。

with the public sphere in the urban environment." Thus, each of these schools has its peculiar way to relate to the city or to the natural landscpape: Copenhagen International School Nordhavn (p.132) stands out on a harbor pier, with terraces and large atriums facing the water, while the Skovbakken School (p.182) plays with its roof lines to link with the residential neighborhood; the Frederiksbjerg School (p.148) lifts up its main volume to create an urban public space below; the Stephen Perse Foundation Senior School (p.166) accommodates an intimate landscaped courtyard and a flexible public space for breakout activities and group study; the German School Madrid (p.196) speaks with the language of concrete, but both in a friendly and inviting manner, via open air or half-covered spaces and courtyards, creating their urban garden.

As for environment and sustainable strategies, it is to be remarked how the schools in Denmark have large inner atriums for all social and physical activities, while the one in Spain conceives these rooms as open air spaces, as the climate consents.

Sustainability should not only be a technical issue worrying the adult and the professional, but addressing the issue to the children will help them develop their own conscience of a sustainable approach in life. The sustainability topic is part of a larger scope we could name "cultural responsibility", including all that a school building can show and offer directly to the children for their evaluation and knowledge enrichment.

Reading by this perspective, the social and cultural responsibilities of the school in modeling the future life of the kids remain central subjects in the design of the modern school building.

The metaphor between the school and the city is only exemplifying the complexity of the design issue. Still, as early as in 1960, Aldo Van Eyck used these words while illustrating the idea behind his Kindertehuis: "A small world inside a big one, a big world in a small house in the city, a city as a house, a house for the children." Highlighting these words is underlining the concept that lays under this selection of schools.

哥本哈根诺德海文国际学校
Copenhagen International School Nordhavn

C.F. Møller Architects

诺德海文国际学校是哥本哈根国际学校的新校舍,位于哥本哈根诺德海文区海滨的一座人造半岛上,周围是大海和待建的绿地。这座2.5万平方米的校舍是哥本哈根最大的学校,可容纳1200名学生和280名教职员工。

这座现代教育建筑旨在将学校建筑与城市环境中的公共空间联系起来,给学校营造一个开放的氛围。学校外的长廊将成为一个城市港口空间,提供放松和举办各种活动的机会。

主教学楼被细分成四座较小的"塔楼",分别是小学、初中和高中,楼高从5层到7层不等,每一层都经过特别的改造,以满足儿童在不同发展阶段的需要。例如,小学生的教室特别宽敞;教室内及其周围都将安排各种各样的活动,每个活动都有指定的绿地,并设有戏剧/表演设施、体育设施等。学校分成四个单元,也为社区活动提供了方便,建立了品牌形象,也更利于查找路径。

大多数教室都被安排在建筑的角落,充分利用从各个侧面入射的阳光,也保证了视野畅通无阻。这些教室均配备了大型的现代化高分辨率交互式白板,空间大,天花板也高。

学校的所有四个单元都建在一层的基座上,在基座层可以举办更常见的、外向型活动,这里包括一个门厅、体育设施、一个食堂、一个图书馆和表演设施。因此,课室可以在正常上课时间之外关闭,而公共区域将继续开放,供学校和当地社区举办各种活动。

基座包括一个公共屋顶露台,它将作为整个学校的操场,尤其是为小学生提供活动空间。架高的学校操场提供了一个安全的环境,防止学生离水太近或远离学校范围。

所有的室内设计都采用了天然材料:地板是涂油橡木,沙发是羊毛软垫的,书架是纯竹子制作的。从长远考虑,学校的卫生间将采用循环水冲洗。

学校建筑独特的外立面覆盖着12000块太阳能电池板,每个电池板都有各自的角度,形成了类似亮片的效果,满足了学校超过一半的年用电量。太阳能电池总占地面积为6048m²,是丹麦最大的建筑一体化太阳能电站之一,预计年发电量超过200MWh。

除了有助于提升学校的绿色形象之外,太阳能电池还构成了学校课程的永久组成部分,学生也可以学习监控能源生产,并在物理和数学课堂上使用相关数据。

CIS Nordhavn is a new school building for the Copenhagen International School, located on the waterfront in Copenhagen's Nordhavn quarter, on a man-made peninsula - surrounded by the sea and coming green areas. The 25,000m² school building is Copenhagen's largest school, and accommodates 1,200 students and 280 employees.
The modern educational architecture is designed to link the school premises with the public sphere in the urban environment, and give the school an open ambience. The promenade outside the school will become an urban port-side space providing opportunities for relaxation and various activities.

The main school building is subdivided into four smaller "towers" – Early Years, Primary School, Middle School and High School – ranging from five to seven storeys, each specially adapted to meet the needs of children at different stages of development. For example, the classrooms for the youngest pupils are particularly large: A full range of functions will take place in and around the classroom, each of which has designated green spaces and areas with drama/performance facilities, PE, etc. The subdivision of the school into four units facilitates community, identity and easy wayfinding.
Most of the classrooms are placed in the building's cor-

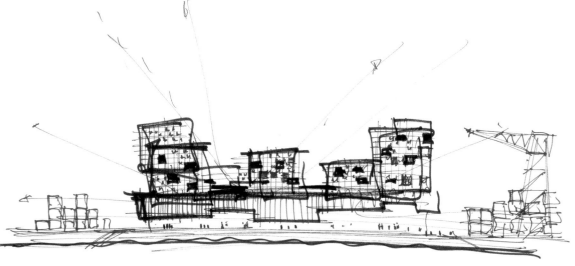

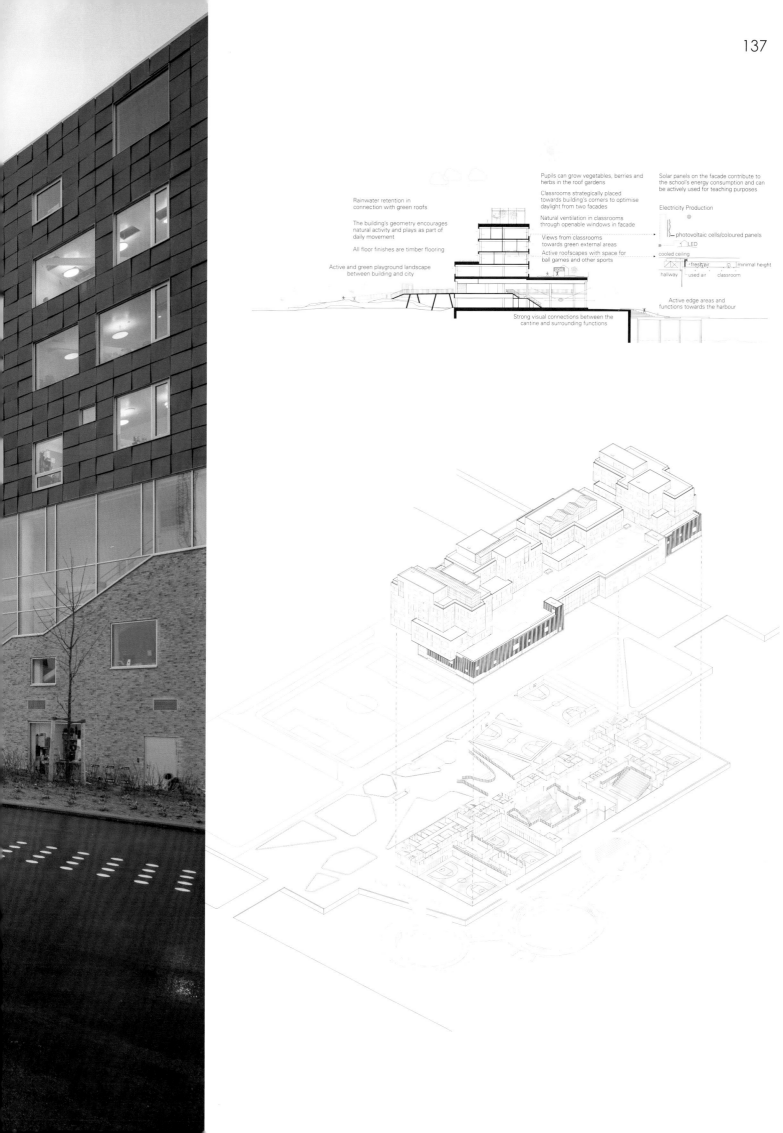

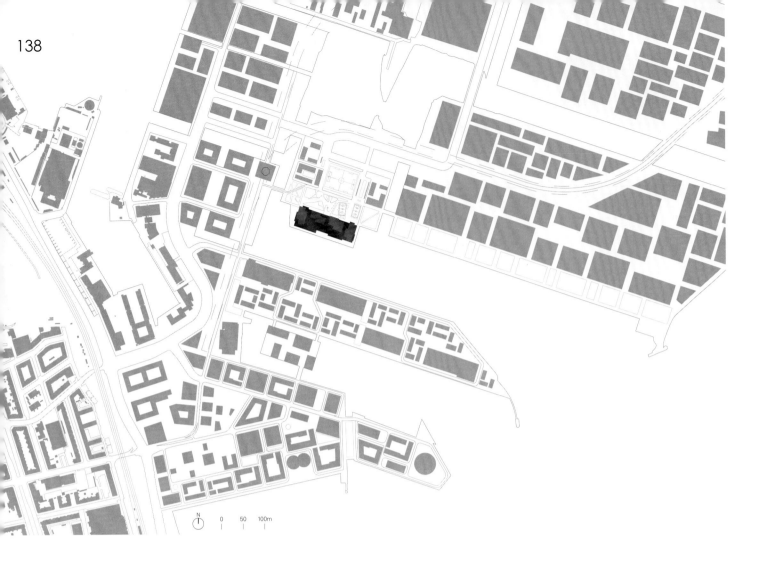

北立面 north elevation

南立面 south elevation

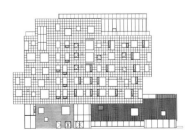
西立面 west elevation

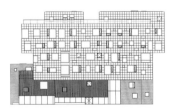
东立面 east elevation

1. 高中
2. 初中
3. 小学
4. 学前教育
5. 教室
6. 办公室
7. 露台
8. 屋顶操场

1. high school
2. middle school
3. primary school
4. early years
5. classroom
6. offices
7. terrace
8. rooftop playground

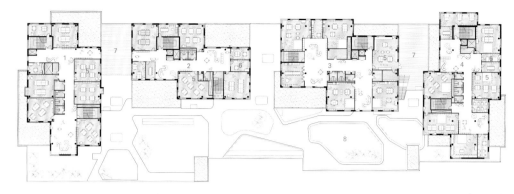

四层 third floor

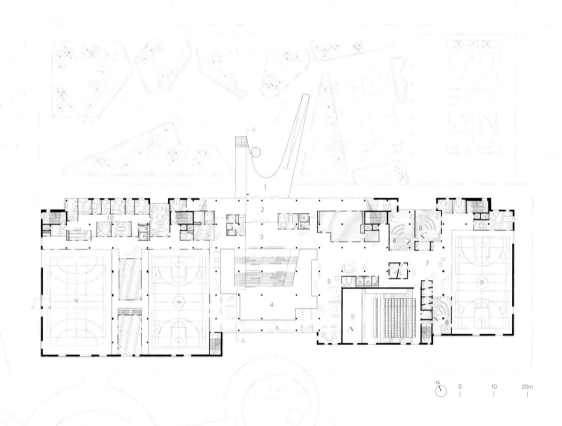

二层 first floor

1. 坡道
2. 主入口
3. 接待处
4. 主大厅
5. 图书馆
6. 音乐教室
7. 休息室
8. 表演空间
9. 体育馆
10. 行政管理区
11. 食堂
12. 学习厨房
13. 设计技术室
14. 更衣室
15. 露台

1. ramp
2. main entrance
3. reception
4. main hall
5. library
6. music room
7. lounge
8. performance space
9. sports hall
10. administration
11. canteen
12. learning kitchens
13. design tech room
14. changing room
15. terrace

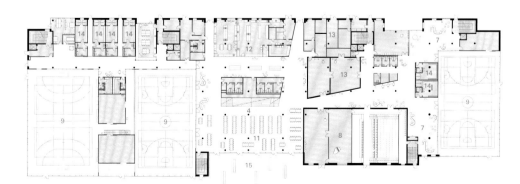

一层 ground floor

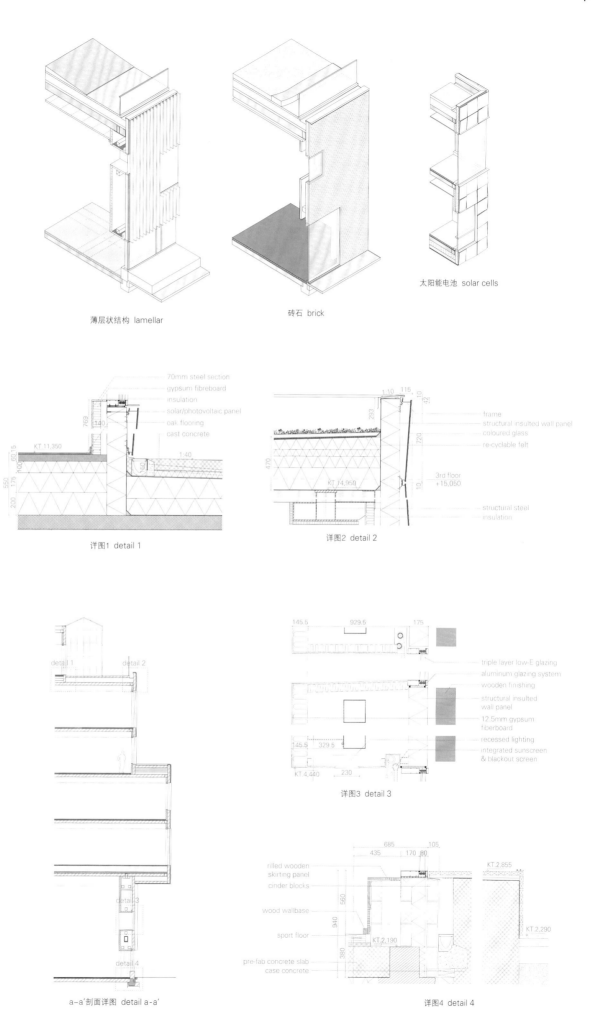

ners, to optimise daylight and views from several sides. The classrooms are equipped with large modern, interactive high-resolution smartboards, and they have plenty of space, with high ceilings.

All four school units are built on top of the ground-floor base, which contains common and more extrovert activities, including a foyer, sports facilities, a canteen, a library and performance facilities. The classroom units can thus be closed-off outside normal school hours, while the common areas will remain open for various events of the school and local community.

The base includes a common roof terrace which will function as a school playground for the whole school – and the youngest pupils in particular. The elevated school playground provides a secure environment, which prevents students from coming too close to the water or from straying off the school premises.

All interiors are in natural materials: the floors are of oiled oakwood, sofas are upholstered in wool, and bookshelves are solid bamboo. In the longer term the school's toilets will be flushed with recirculated water.

The school building's unique facade is covered in 12,000 solar panels, each individually angled to create a sequin-like effect, which supply more than half of the school's annual electricity consumption. The solar cells cover a total area of 6,048 square meters, making it one of the largest building-integrated solar power plants in Denmark, estimated to produce over 200 MWh per year.

In addition to contributing to the school's green profile, the solar cells also form a permanent part of the school's curriculum, allowing students to monitor energy production and use data in physics and mathematics classes.

A-A'剖面图 section A-A'

项目名称：Copenhagen International School Nordhavn / 地点：Levantkaj, Nordhavnen, Copenhagen, Denmark / 建筑师：C.F. Møller Architects
项目团队：Mads Mandrup – partner in charge, Jørgen Juul, Thue Borgen Hasløv, Anne Lilke Krag Hansen, Peter Michael Dolf, David Vega, Claus Christian Tversted, Mihaila Periklieva Terzis-Georgieva, Henrik Vedel Larsen, Steen Kortbæk Svendsen, Ida Richter Brændstrup, Simon Olafsson, Mikkel Daniel Sørensen, Frans Borgmann, Nina Walsh, Mette Lyng Hansen / 景观设计师：C.F. Møller Landscape / 工程师：NIRAS A/S / 主要承包商：Per Aarsleff A/S, Eiler Thomsen Alufacader, Consortium KT DTEK. / 客户：Ejendomsfonden Copenhagen International School (ECIS) / 用地面积：16,500m² / 建筑面积：6,000m² / 总建筑面积：26,000m² (accommodating 1,200 students) / 结构：steel pre-fab, concrete pre-fab / 材料：solar cell cladding, brick, timber cladding, wood/bamboo flooring + furnishing / 室内材料：floors - oiled oakwood; sofas - upholstered in wool; bookshelves - solid bamboo / 造价：DKK 500 million / 项目时间：2013—2017 / 摄影师：©Adam Mørk (courtesy of the architect)

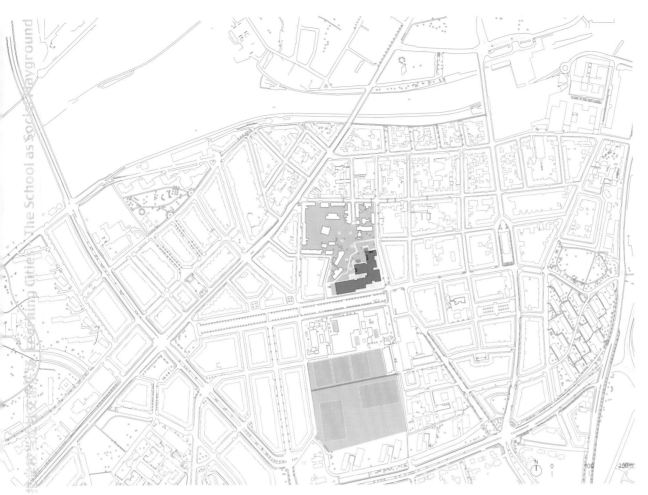

丹麦 Frederiksbjerg 学校

Frederiksbjerg School

Henning Larsen Architects + GPP Arkitekter

Frederiksbjerg学校是近一百年来丹麦第二大城市奥尔胡斯中部的第一所新建学校。学校位于Frederiksbjerg区，主要特色是四至六层高的红砖建筑。

学校的主入口位于两条历史悠久的街道的拐角处，两层高架空结构下方支撑了混凝土柱子，与中庭的柱子相匹配。一条宽大的南向楼梯沿着林荫大道而建，通常用作长椅，将学校与周围环境融合在一起。

学校围绕着一个中心中庭组织，建筑的各个部分在这里相聚并互相连接。共享的实习教室、实验室和学习厨房与中庭相连。当你从主入口穿过建筑时，从开放的中庭开始，穿过较小的共享区域，再进入集中在一起的小教室和小组研讨室。这种结构设计在所有的三个楼层都是一样的。日托中心、年龄最小的孩子和行政人员都安排在一楼，初中学生在二楼，再大一点的学生则在三楼。学生和老师在中庭见面，中庭是楼层之间的垂直开放连接，这样的设计旨在增强学生和教师之间的视觉和实际联系。

大量的教室围绕一个共享的中心房间而建，鼓励举办各种活动，拓展不同的研究领域。活动区域可以让学生通过运动和玩耍来集中精力学习。这些区域特别适合不同年龄组的孩子们，他们理解能力和运动水平都不一样。学习区被打造成小壁龛的模式，为个人学习打造了安静的空间。

教室位于建筑集群当中，每间教室都有自己的小组研讨室，从教室和共享区域两侧都能进入。课余时间，学校的教室、运动场和体育馆面向公众和当地的体育协会开放。

学校与周围住宅和公共机构共享公共操场和室外区域。该区域有一个户外厨房、工具棚、兔子窝和一个迷你月球车库。

立面由再生砖制成。事实上，20%的覆层材料取自原先的St. Annagades学校。古铜色的砖使建筑与周围的历史环境产生了关联。砖块中也包含了以前建筑的经历，为新学校增添了强烈的认同感。这种有手感的红色材料为建筑外观增添了热情和温暖的感觉，此外，混凝土支柱、黑色和无烟煤灰色表面以及彩色涂料也为建筑外观增添了活力。

建筑内部的颜色和材料有利于确定方位，也便于提供后勤服务。

南立面 south elevation

两个大厅几乎漂浮在中庭之中,大厅表面覆盖着红色和橙色的吸声板,三幅宏伟的壁画环绕着独立的小组研讨室。

　　Frederiksbjerg学校的设计将日光作为一种不断变化的光源。在建筑的设计中,窗户的大小是分级的,最大的窗户处于立面的中心,较小的窗户在顶部的屋顶附近,最小的窗户则接近地面。窗户的排布方式在新建筑的日光体验中创造了自然的多样性。

Frederiksbjerg School is the first new-built school located in the southern part of the central Aarhus, the second largest city in Denmark, in approximately 100 years. The school is placed in the Frederiksbjerg district where it is mostly characterized by 4-6 story building blocks in red bricks.
The main entrance of the school is placed on the corner of two historic streets and a cantilever is built in double height with concrete pillars which match the pillars in the atrium. Built along the boulevard is a big, south-facing staircase, often used as benches, which merges the school with the surrounding context.
The school is organized around a central atrium where clusters of the building meet and join together. Shared practical rooms, laboratories and learning kitchens are located in connection with the atrium. When you move through the building from the main entrance, you start at the open center atrium through the smaller shared areas to the small class and group rooms in the clusters. This structure is repeated on all three floors. The daycare, the youngest children and the administration are placed on the ground

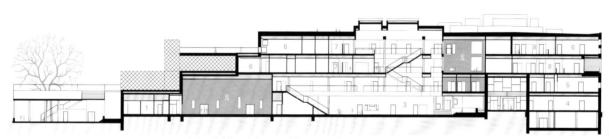

1. 更衣室 2. 体育馆 3. 咖啡厅 4. 行政管理区 5. 保健室 6. 教室 7. 更衣室 8. 小教室
1. changing room 2. gym 3. cafe 4. administration 5. health 6. classroom 7. wardrobe 8. cluster
A-A'剖面图 section A-A'

1. 小组研讨设施 2. 工作坊 3. 教室 4. 图书馆 5. 音响室 6. 更衣室 7. 小教室 8. 体育馆 9. 学习室
1. group facility 2. workshop 3. classroom 4. library 5. audio 6. wardrobe 7. cluster 8. gym 9. study
B-B'剖面图 section B-B'

1. 教室 2. 室外工作坊 3. 创意工作坊 4. 图书馆 5. 小教室
1. classroom 2. outdoor workshop 3. creative workshop 4. library 5. cluster
C-C'剖面图 section C-C'

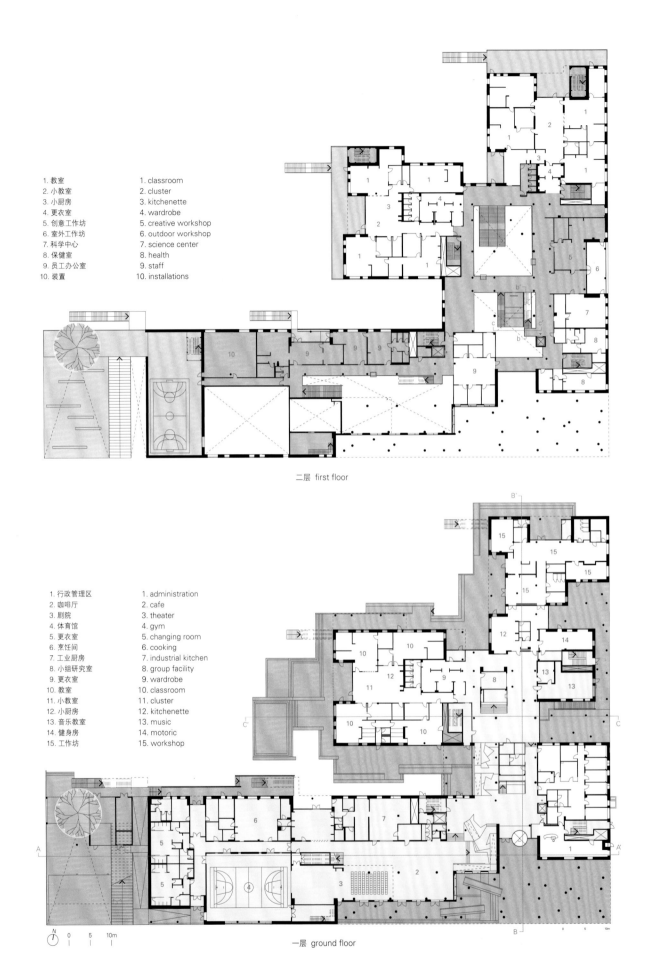

1. 教室	1. classroom
2. 小教室	2. cluster
3. 小厨房	3. kitchenette
4. 更衣室	4. wardrobe
5. 创意工作坊	5. creative workshop
6. 室外工作坊	6. outdoor workshop
7. 科学中心	7. science center
8. 保健室	8. health
9. 员工办公室	9. staff
10. 装置	10. installations

二层 first floor

1. 行政管理区	1. administration
2. 咖啡厅	2. cafe
3. 剧院	3. theater
4. 体育馆	4. gym
5. 更衣室	5. changing room
6. 烹饪间	6. cooking
7. 工业厨房	7. industrial kitchen
8. 小组研究室	8. group facility
9. 更衣室	9. wardrobe
10. 教室	10. classroom
11. 小教室	11. cluster
12. 小厨房	12. kitchenette
13. 音乐教室	13. music
14. 健身房	14. motoric
15. 工作坊	15. workshop

一层 ground floor

项目名称：Frederiksbjerg School / 地点：Aarhus, Denmark / 建筑师：Henning Larsen Architects + GPP Arkitekter
景观设计师：Møller & Grønborg / 总承包商：Hoffmann / 工程师：Niras / 可持续设计：Building Class 2020 (according to the Danish Building Regulations 2010) /
项目团队：Peer Teglgaard Jeppesen (responsible partner), Margrete Grøn (project manager), Anders Nielsen, Dorte Nielsen, Eva Bryzek, Eva Ravnborg,
Henrik Jacobsen, Peter Munch, Vanda Oliveira, Vanja Scott, Zazia Wihlborg Bigom, Tobias Dræger, Kasper Christiansen, Glenn Collett Poulsen – Henning Larsen Architects

1. 教室
2. 小教室
3. 小厨房
4. 更衣室
5. 学习室
6. 音乐教室
7. 民主大厅
8. 体育馆
9. 科学实验室
10. 室外工作坊

1. classroom
2. cluster
3. kitchenette
4. wardrobe
5. study
6. music
7. democracy hall
8. gym
9. science laboratory
10. outdoor workshop

四层 third floor

1. 教室
2. 小教室
3. 小厨房
4. 更衣室
5. 图书室
6. 音响室
7. 音乐教室
8. 科学室
9. 体育馆
10. 实践工作坊
11. 手工工作坊
12. 创意工作坊
13. 员工办公室

1. classroom
2. cluster
3. kitchenette
4. wardrobe
5. library
6. audio
7. music
8. science
9. gym
10. practical workshop
11. hand workshop
12. creative workshop
13. staff

三层 second floor

平面设计、指路标示：Bodil Nordstrøm – Lead graphic designer; Andreas Engelbreckt Bünger, Sebastian Simonsen / 能源与日照计算、材料策略：Jakob Strømann-Andersen, Anne Iversen, Martha Lewis, Erik Folke Holm-Hansson, Micki Aaen Petersen / 客户：Aarhus City Council / 总建筑面积：15,000m²
施工时间：2014—2016 / 摄影师：©Hufton + Crow (courtesy of the architect) - p.148~149, p.150~151, p.153, p.154, p.157, p.160~161, p.163, p.164~165; ©Peter Nørby (courtesy of the architect) - p.152, p.155 left-top, middle, bottom, p.156, p.162; ©Virklund Sport (courtesy of the architect) - p.155 right-top, middle, bottom

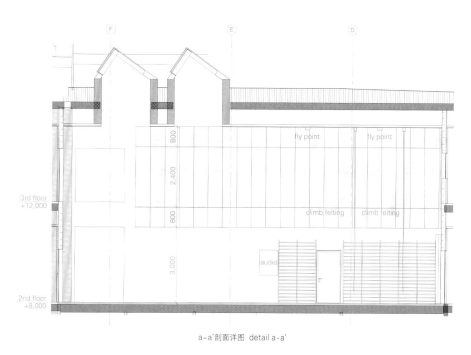

a-a'剖面详图 detail a-a'

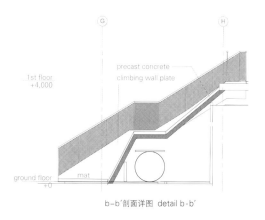

b-b'剖面详图 detail b-b'

c-c'剖面详图 detail c-c'

and the first floors, the middle grades students are on the second floor and the oldest students are on the third floor. The students and the teachers meet in the atrium which is the vertical open connection between the floors. This targets the ambition to strengthen the visual and physical connection between students and teachers.

The clusters are built around a shared center room encouraging various activities and study areas. The activity areas are used for the activity that focuses on learning through movement and play. These areas are specifically fitted to the different age groups and levels of understanding and motion. The study areas are built as small niches which create quiet rooms for individual study.

The classrooms are located in the clusters of the building and each classroom has its own group room, which can be accessed from both classrooms and from shared areas. Outside school hours, the classrooms, playing fields and sports halls are open to the public and the local sporting associations.

The school shares public playgrounds and outside areas with the surrounding houses and institutions. The area has an outdoor kitchen, tool sheds, rabbit hutches and a mini

moon car garage.

The facades are made of recycled bricks. In fact, 20 percent of the cladding stems from the old St. Annagades School. The patinated bricks give the building a contextual relation to the historical surroundings. The brick also tells the stories of former buildings adding a strong identity to the new school. The tactile, red material adds a warm and welcoming expression to the exterior, which furthermore appears with pillars of concrete, black and anthracite grey surfaces and colored coatings.

Inside the building the colors and materials support the orientation and logistics of the school. Two halls which almost float in the atrium are covered with red and orange acoustic tiles and three big wall paintings wrap themselves around the detached group rooms.

The design of Frederiksbjerg School focuses on daylight as an ever-changing light source. In the design of the building the sizes of the windows have been graded, being largest in the center of the facade, smaller in the top by the roof and smallest near the ground. The pattern of the window creates a natural diversity in the experience of the daylight in the new building.

斯蒂芬·佩尔斯基金会高中
Stephen Perse Foundation Senior School
Chadwickdyerclarke Studio

该项目由位于剑桥市中心一座新建成的五层高开发项目组成,项目位于斯蒂芬·佩尔斯基金会高中部场地北部边缘的新城和格里森路保护区内。该新建筑容纳了围绕新的体育和娱乐空间布置的一些教学空间。

屋顶的运动场地和一个有四个场馆的体育馆被精心整合到一个建筑体量中。设计开发中的一个关键问题就是考虑到了周围道路的环境和特点,附近既有普通住宅,也有规模更大的大学教学楼。

新建筑位于两栋现有的教学楼之间,并与之相连,改善了两者的通达性,为整个高中场地提供了"连接件"。向西,在新旧交汇处形成了新的双层通高入口和接待空间。从学校的"珍奇柜"图书馆可以进入一个私密的庭院,这里为学习和放松提供了安静的空间。

设计的驱动力是为学生提供一系列外部空间,并将学校场地和住宿需求结合在一起。建筑师与景观设计合作伙伴制定了清晰的景观策略,充分利用了场地以前的未利用区域,释放了许多空间供学生使用。景观设计在学校的中心提供了一个有吸引力的环境,同时也能支持外部教学、学习和体育活动。

体育馆位于地下一层,设有更衣室等配套设施。这样就减少了建筑的体量,但也使得这个宽敞的新房间成为新交通流线系统的中心,从学校的所有位置都可以看到。在一层,一个灵活的活动空间形成了学校的多功能心脏,它向周围的景观开放,在双层高的雨篷下,由梁和扭曲的柱子编织结构支撑。这个雨篷在一定程度上支撑着垂直跨越体育馆的屋顶运动场,为学校提供了一个令人振奋的外部遮挡。

建筑的上面两层容纳了围绕双层高"中心"布置的许多教学空间,旨在形成分组活动和小组学习的焦点。该建筑采用交叉层压木材建造,这是一种低能耗的解决方案,这些木材是在现场之外预制的,以方便在狭窄的场地上安装。建筑呈开放的状态,增强了低能耗和自然通风策略。

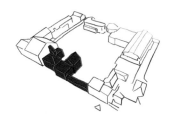 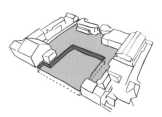 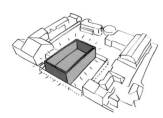 

北立面 north elevation

西立面 west elevation

The project comprises the development of a new five storey building in the central Cambridge, sitting within the New Town and Glisson Road conservation area on the northern edge of the Stephen Perse Foundation's Senior School site. The building houses a number of teaching and learning spaces arranged around new sports and recreation spaces. A rooftop sports pitch and a four-court sports hall have been carefully integrated into a contextual piece of architecture that addresses the volumes of these active spaces. Consideration of the context and character of the surrounding roads were key issues in the design development, given the proximity of both domestic properties and larger university buildings.

The new building sits between - and connects to - two existing school buildings, improving access to both and providing "connective tissue" for the senior school site as a whole. To the west, a new double-height entrance and reception space are formed at the junction of new and old. An intimate landscaped courtyard accessible from the school's "cabinet of curiosity" library provides a quiet space

项目名称：Stephen Perse Foundation Senior School / 地点：Cambridge, UK / 建筑师：Chadwickdryerclarke Studio / 项目管理：Bidwells / 规划顾问：Beacon Planning / 结构工程师：Smith and Wallwork / 机电设备工程师：Mott MacDonald / 声效设计：Sharps Redmore / 景观设计：The Landscape Partnership / 造价管理和CDM顾问：Aecom / 总承包商：Kier Eastern / 客户：The Stephen Perse Foundation / 用地面积：3,700m² / 总建筑面积：2,940m² / 建筑占地面积：1,320m² / 施工造价：Approx £3,250 per sqm excl. VAT & fees / 项目造价：£9.4 million / 规划许可获得时间：2015.11 / 设计时间：2017.9 / 摄影师：©Richard Chivers (courtesy of the architect)

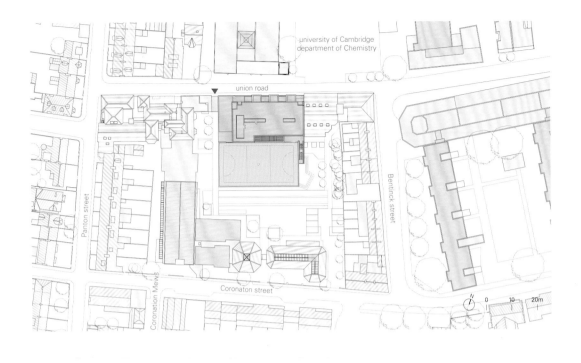

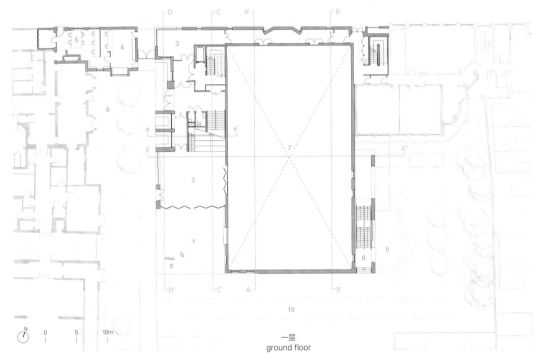

1. 室外雨篷
2. 活动空间
3. 中心空间
4. 接待处
5. 办公室
6. 通往多用途游戏区的室外楼梯
7. 体育馆上方的挑空空间
8. 图书馆庭院
9. 舞台区域
10. 学校操场

1. external canopy
2. activity space
3. hub space
4. reception
5. office
6. external stair to MUGA (multi use games area)
7. void over sports hall
8. library courtyard
9. stage area
10. school play area

一层
ground floor

1. 体育馆
2. 更衣室
3. 员工更衣室
4. 健身房
5. 商店
6. 设备间
7. PE办公室
8. 推拉式看台座椅

1. sports hall
2. changing rooms
3. staff changing
4. gym
5. store
6. plant
7. PE office
8. pull out bleacher seating

地下一层
first floor below ground

1. 教室
2. 中心空间
3. 室外露台
4. 多用途游戏区的屋顶挑空
5. 设备间

1. classroom
2. hub space
3. external terrace
4. rooftop MUGA (void)
5. plant room

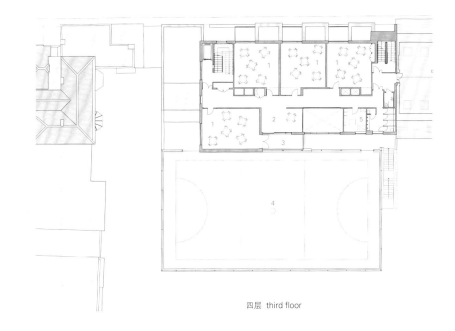

四层 third floor

1. 员工区
2. 教室
3. 中心空间
4. 室外露台
5. 多用途游戏区的屋顶挑空

1. staff area
2. classroom
3. hub space
4. external terrace
5. rooftop MUGA

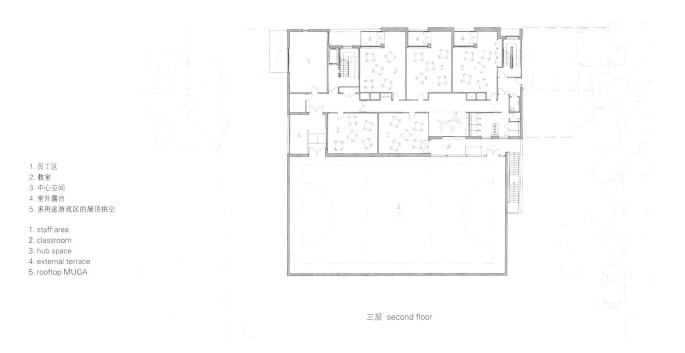

三层 second floor

1. 创意室
2. 活动空间
3. 静音室/会议室
4. 办公室
5. 会议室
6. 体育馆上方的挑空空间
7. 通往多用途游戏区的室外楼梯
8. 连桥
9. 室外雨篷

1. innovation room
2. activity space
3. quiet room/meeting room
4. office
5. meeting room
6. void over sports hall
7. external stair to MUGA
   (multi use games area)
8. bridge link
9. external canopy

二层 first floor

for study and relaxation.

The design has been driven by the need to provide a sequence of external spaces for the use of the pupils and knit the school site together as much as its accommodation requirements. A clear landscape strategy developed with The Landscape Partnership intensifies and liberates previously underused areas of the site for pupil use. The landscape provides an attractive environment at the heart of the school whilst supporting external teaching, learning and physical activity.

The sports hall is located at basement level, along with changing rooms and other support facilities. This has allowed the mass of the building to be reduced, but also allow this large new room to be central to a new pattern of circulation and be visible from all points of the school.

At the ground floor, a flexible activity space forms a multi-functional heart to the school which opens up to the surrounding landscape and beneath a double-height canopy, supported by a woven structure of beams and twisting columns. This canopy partly supports the rooftop sports pitch, which spans perpendicularly across the sports hall, and provides an exciting external shelter for the school. The upper two storeys of the building house a number of teaching and learning spaces arranged around a double-height "hub" intended to form a focus for breakout activities and group study. The building is constructed using the cross laminated timber, which is a low-embodied energy solution and pre-fabricated off site to enable ease of erection on a constricted site. This is broadly exposed and augments the low-energy and natural ventilation strategies.

1. 体育馆
2. 推拉式看台座椅
3. 交通流线空间
4. 通往体育馆的交通流线空间/观景走廊
5. 中心空间
6. 教室
7. 室外露台
8. 多用途游戏区的屋顶
9. 通风井
10. 屋顶种植区

1. sports hall
2. pull out bleacher seating
3. circulation
4. circulation / viewing gallery to sports hall
5. hub space
6. classroom
7. external terrace
8. rooftop MUGA
9. ventilation shaft
10. roof plant space

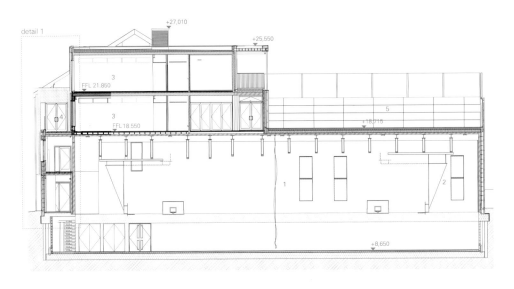

1. 体育馆
2. 推拉式看台座椅
3. 教室
4. 室外露台
5. 多用途游戏区的屋顶

1. sports hall
2. pull out bleacher seating
3. classroom
4. external terrace
5. rooftop MUGA

B-B'剖面图 section B-B'

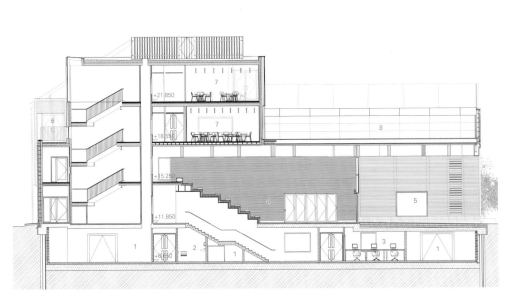

1. 商店
2. 更衣室
3. PE办公室
4. 活动区
5. 室外雨篷
6. 室外露台
7. 教室
8. 多用途游戏区的屋顶

1. store
2. changing rooms
3. PE office
4. activity space
5. external canopy
6. external terrace
7. classroom
8. rooftop MUGA

C-C'剖面图 section C-C'

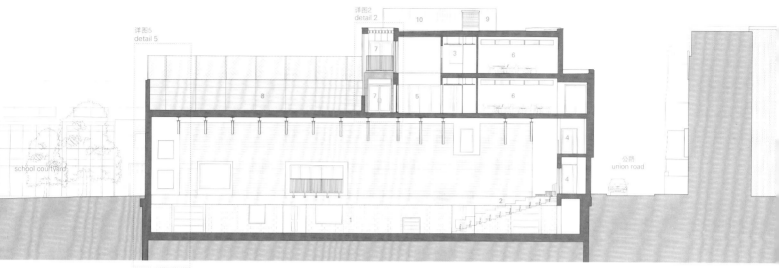

A-A'剖面图 section A-A'

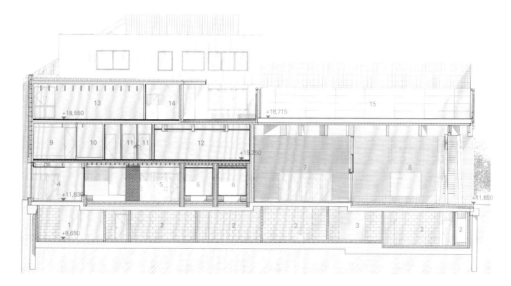

1. 设备间
2. 更衣室
3. 健身房
4. 中心
5. 主入口
6. 室外阅读区
7. 活动区
8. 室外雨篷区
9. 静音室/会议室
10. 通往连桥的坡道
11. 卫生间
12. 创意室
13. 员工区
14. IT部门
15. 多用途游戏区的屋顶

1. plant
2. changing room
3. gym
4. hub
5. main entrance
6. external reading booths
7. activity space
8. external canopy space
9. quiet room / meeting room
10. ramp to bridge
11. wc
12. innovation room
13. staff area
14. IT patch room
15. rooftop MUGA

D-D'剖面图 section D-D'

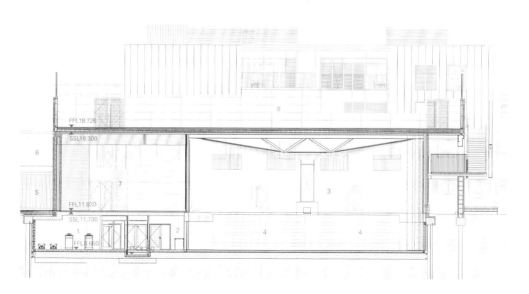

1. 健身房
2. PE办公室
3. 体育馆
4. 推拉式看台座椅
5. 入口大门
6. 后方的连桥
7. 活动空间
8. 多用途游戏区的屋顶

1. gym
2. PE office
3. sports hall
4. bleacher seating
5. entrance gate
6. bridge link behind
7. activity space
8. rooftop MUGA

E-E'剖面图 section E-E'

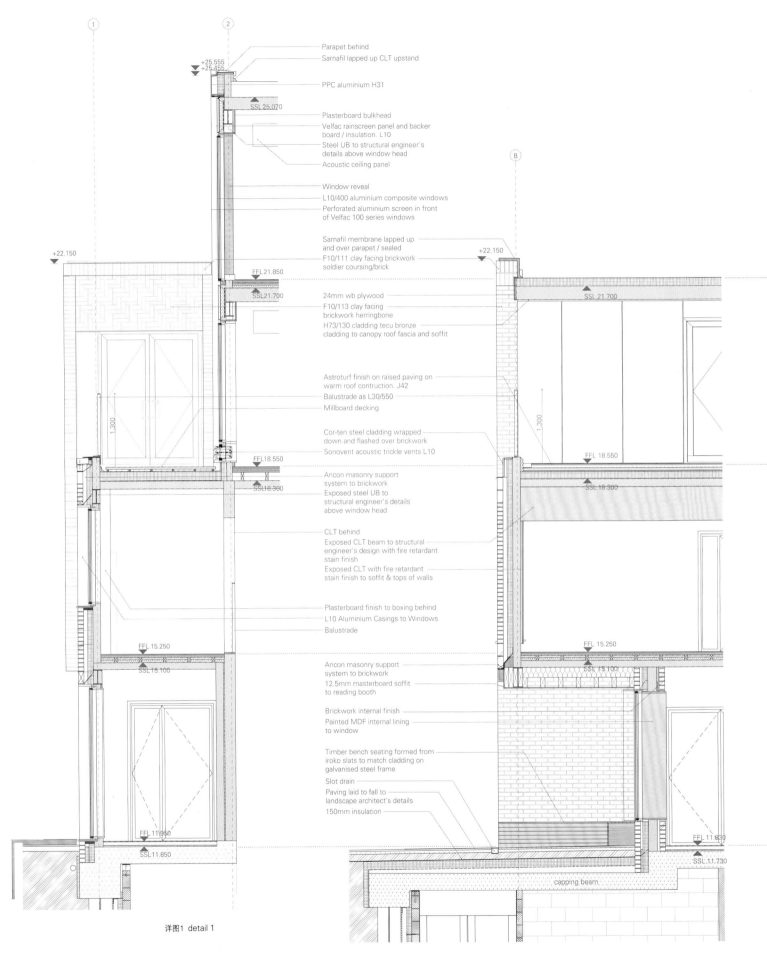

详图1 detail 1

a-a' 剖面详图 detail a-a'

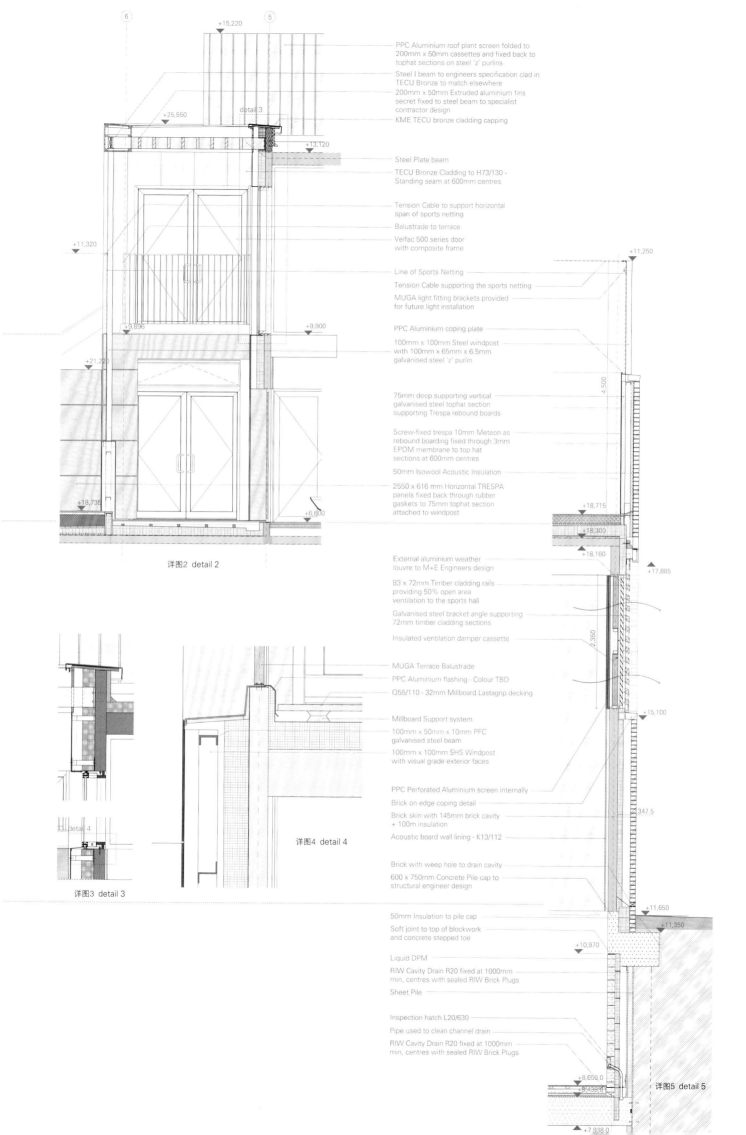

斯科夫巴肯学校
Skovbakken School

CEBRA

在丹麦奥尔胡斯附近奥德市的斯科夫巴肯，学生和老师们正在张开双臂欢迎他们的新学校。这所两层楼的学校拥有独特的坡屋顶，是奥德市对新建筑的最大投资，表明了支持现代教育和教学的决心。

斯科夫巴肯学校占地9300m²，共两层，包括小学、初中和高中，设有教室、体育馆和日托设施。新学校的建筑表达方式和所选用的材料均受到了周围社区和森林的启发。

在建筑场地尽可能多地保留了许多原始树木。那些必须砍伐的树则被用来装饰室内。木质部分突出了所有的入口；在建筑内部，引人注目的楼梯也是由木材制成的。木材的使用保持了与建筑风格的实际联系。学校的布局更是突出了热情好客的特点。三个结构体围绕一个中庭，打通周围环境，在室外和室内创造迷人的景观空间。

学校把锻炼身体当成了日常生活的一部分。例如，公共区域的消防通道也是赛道。每个教室都有指定的区域进行体育活动。公共休息室直接通往健身房，健身房可以在休息时用作游戏空间。

在设计新斯科夫巴肯学校时，建筑师的任务是替换一系列旧建筑，这些建筑已经扩建了好几次，因此看起来像是拼凑而成的，并不适

合现代教学。新的综合体结合了学校和日托设施，在一个通透而又多样化的新学习环境中，鼓励学科领域之间的互动。此外，它也使体育活动和玩耍成为学校生活的一部分。

　　高低天花板、光线明亮和昏暗、小空间和大空间的组合，让孩子们能够根据自己的需要和心情参与不同的社交场合——大型集会、小型团体活动或独处。此外，建筑的平面布局留出了一系列的户外空间，可以适应不同的需求：晒太阳、遮阳和遮风挡雨。这些多功能空间可以根据学校日常生活的层次和活动模式的变化来使用和配置。

At Skovbakken in Odder, near Aarhus in Denmark, students and teachers welcome their new school. The largest investment in a new building in Odder Municipality, the two-storey school with its distinctive pitched roof supports modern education and teaching.

The Skovbakken School covers 9,300m² over two floors housing a primary, middle and senior school with classrooms, a sports hall and daycare facility. The architectonic expression and materials of the new school are inspired by the surrounding neighbourhood and forest.

The site preserved many of the original trees as possible. The trees that had to be cut down were used to furnish the interior. Wooden sections accentuate all entrances; inside, the striking staircase is also made of wood. The use of wood maintains physical communication with the architecture.

The welcoming character of the school is emphasised in the school's layout. Three structures surround a central atrium, opening up the surroundings, creating intriguing landscaping spaces outside and building spaces inside.

The school makes exercise part of everyday life. Thus, for

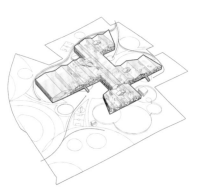

1. 特殊教室　　1. specialised classroom
2. 学习区　　　2. study area
3. 衣帽间　　　3. locker room
4. 教室　　　　4. classroom
5. 小组学习室　5. group study room
6. 图书室　　　6. library
7. 储藏室　　　7. store room
8. 教工办公室　8. staff room
9. 行政管理区　9. administration
10. 接待处　　10. reception

二层 first floor

1. 特殊教室　　9. 舞台
2. 学习区　　10. 自助餐厅
3. 衣帽间　　11. 主入口
4. 教室　　　12. 行政管理区
5. 小组学习室 13. 厨房
6. 体育馆　　14. 日托中心
7. 健身房　　15. 婴儿床房间
8. 更衣室

1. specialised classroom
2. study area
3. locker room
4. classroom
5. group study room
6. sports hall
7. gym
8. changing room
9. stage
10. cafeteria
11. main entrance
12. administration
13. kitchen
14. daycare center
15. crib room

一层 ground floor

立面1 elevation 1　　立面2 elevation 2　　立面3 elevation 3

A-A'剖面图 section A-A'　　立面7 elevation 7

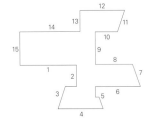

立面12 elevation 12　　立面13 elevation 13

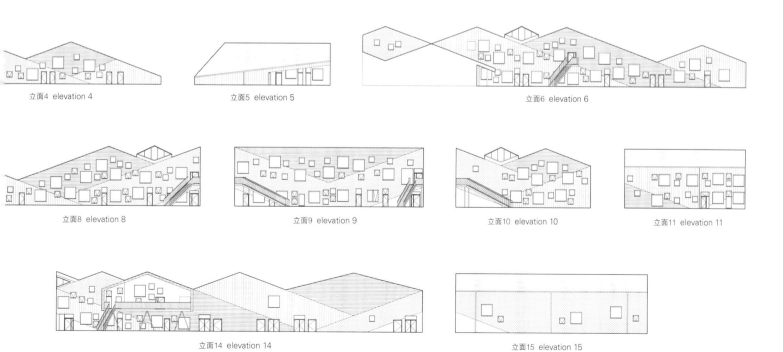

立面4 elevation 4　　立面5 elevation 5　　立面6 elevation 6

立面8 elevation 8　　立面9 elevation 9　　立面10 elevation 10　　立面11 elevation 11

立面14 elevation 14　　立面15 elevation 15

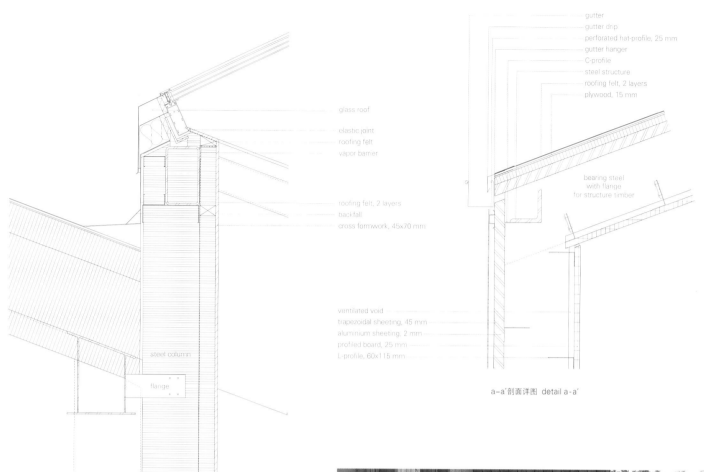

详图1 detail 1

- glass roof
- elastic joint
- roofing felt
- vapor barrier
- roofing felt, 2 layers
- backfall
- cross formwork, 45x70 mm
- steel column
- flange
- ventilated void
- trapezoidal sheeting, 45 mm
- aluminium sheeting, 2 mm
- profiled board, 25 mm
- L-profile, 60x115 mm

- gutter
- gutter drip
- perforated hat-profile, 25 mm
- gutter hanger
- C-profile
- steel structure
- roofing felt, 2 layers
- plywood, 15 mm
- bearing steel with flange for structure timber

a-a'剖面详图 detail a-a'

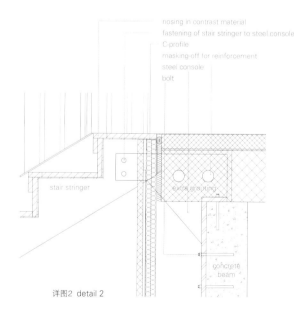

详图2 detail 2

- nosing in contrast material
- fastening of stair stringer to steel console
- C-profile
- masking-off for reinforcement
- steel console
- bolt
- stair stringer
- extra grouting
- concrete beam

example, the fire access routes in the common areas are also tracks. Each classroom has a designated area for physical activity. The common room has direct access to the gym, which can be used as a play space during breaks.

The architect's task with the new Skovbakken School was to replace a series of old buildings that had been expanded several times and therefore looked like a patchwork, ill-suited for modern teaching. The new complex combines a school and a daycare in a diverse and transparent new learning environment that encourages interaction between subject areas. In addition, it allows physical activity and games to be a natural part of the school day.

The combination of high- and low-ceilinged, light and dim, small and large spaces allows the children to engage in different social situations – large assemblies, smaller groups or alone – depending on their needs and moods. In addition, the layout allows for a range of outdoor spaces that can accommodate varying needs for sunlight, shade and shelter. These versatile spaces can be used and equipped in response to the changes in the levels and patterns of activity that define daily life at the school.

项目名称：Skovbakken School / 地点：Tværgade 12, 8300 Odder, Denmark
建筑师：CEBRA / 合作方：MT Højgaard, DEAS
结构工程师：MT Højgaard Design & Engineering
景观设计师：CEBRA / 设备：Sub-consultant – architect and landscape
客户：Odder Municipality / 功能：public school for 650 children, daycare for 100 children / 能量级别：Low energy class standard 2020
用地面积：29,200m² / 建筑面积：9,300m²
结构：lightweight prefabricated facade elements built on steel structure; load-bearing pre-cast concrete walls (primarily those that cut across the facade); longitudinal structural precast concrete slabs (avoiding beams as part of the facade); primary load-bearing walls (transversal walls)
材料：wood (entrance areas, balconies, interior staircases and plateaus), three grey shades (facade cladding), grey colors (the roofing felt), environmentally friendly Troldtekt acoustic ceilings (ceilings)
造价：DKK 130 mio.(€ 17,5 mio.)
项目时间：2015.11—2018.1
摄影师：©Adam Mørk (courtesy of the architect)

马德里德语学校
German School Madrid

Grüntuch Ernst Architects

Learning Cities – The School as Social Playground 学习型城市——把学校当成社交游乐场

马德里北部的这所新学校取代了市中心的旧学校。马德里德语学校以其特殊的课程和晚间的戏剧表演、音乐会,已成为该地重要的文化交流场所。

在这个大型综合体中,学校的区域由易于识别的单元组成。幼儿园、小学和中学都有一个内庭院,为传统修道院肃穆的氛围增添了现代气息。与此同时,天井对着附近的山脉开放,让人可以欣赏到雪山的景色。

考虑到建筑设计要求和场地的地形,建筑师创造了一个具有强烈雕塑感的有机整体,门厅庭院、自助餐厅、750个座位的音乐厅/礼堂和体育馆连接着各座建筑。

早上,孩子们聚集在门厅庭院里,然后分散前往他们位于不同建筑的教室。在这里,多边形天窗的强烈雕塑感打造了一场迷人的光影游戏。

每座建筑都有自己的特点。通透的自助餐厅在视觉上延伸到了门厅庭院和外部空间,而礼堂在其如同雨篷一般的屋顶下方提供了一种更为专注的氛围。

学校建筑内苍白的色调让人联想起附近谢拉山脉的景观,其余部位采用强调色则突出了室内空间。在楼梯间,彩色玻璃过滤了日光,有时会让人想起发光的花朵。

这些建筑既传达了文化,也体现了可持续性。回归传统的简洁设计,同时展示创新的尖端技术,为建筑构造和建筑设备提供信息。古老的做法将确保建筑的可持续运行,例如,通过迷宫一般的地下热力空间进行自然冷却。将建筑安置在柱子上,增加带顶的室外区域,并在立面上做缩进处理,可减少夏季炎热带来的影响,而大型建筑体量可储存冷空气,使室内气候更加舒适。即使学校的建筑密度很高,合适的保温标准和热回收通风系统也能保证较低的采暖和制冷负荷以及良好的空气质量。该设计通过建筑设计和能源概念的紧密互动,对西班牙中部的主要气候条件做出了反应。

这所学校不仅是由教室排列而成的矩阵,更是学生学习环境的核心。正是学校塑造了学生们对建筑和自然世界的理解以及他们的社会文化体验。学校内部的空间构成和视觉联系促进了团体意识和跨文化交流,延续了德语学校的传统和成功模式。

The new school in northern Madrid replaces its predecessor in the city center. With its special curriculum and evening theater performances and concerts, the German School Madrid is an important site for cultural exchange.

In this large complex, the school's areas consist of readily identifiable units. The kindergarten, primary school and secondary school each frame an inner courtyard, lending a contemporary flavour to the traditionally contemplative atmosphere of a cloister. At the same time, the patios open up to the nearby Sierra, revealing a vista of the snow-covered mountains.

With consideration of the requirements of the architectural brief and the topography of the site, the buildings create an organic ensemble with a strong sculptural presence. The foyer courtyards, the cafeteria, a concert hall / auditorium with 750 seats and the sports hall connect the individual buildings.

In the morning, the children gather in the foyer courtyards before dispersing to their classrooms in the different buildings. Here, the sculptural strength of the polygonal skylights creates a captivating play of light and shade.

Each building has its own characteristics. The exposed cafeteria extends visually into the foyer courtyard and the outside spaces beyond, while the auditorium offers a more

南立面 south elevation

景观设计 landscape design

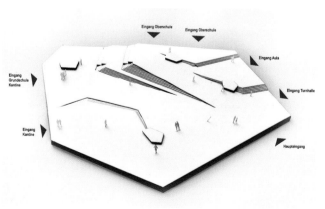

设计研究——门厅庭院的屋顶
design study_roof of the foyer courtyard

门厅庭院地下室：坡道和楼梯一景
basement of the foyer courtyard: a landscape of ramps and stairs

focused ambience beneath its canopy-like roof.
The pale tones inside the school building evoke the hues of the surrounding Sierra mountain landscape. Accent colours highlight the interiors. In the staircases, coloured panes of glass filter the daylight, sometimes recalling luminescent flowers.

The buildings convey both culture and sustainability. A return to traditional simplicity, while demonstrating innovative technological sophistication, informs the construction and building services. Ancient practices, such as natural cooling through a subterranean thermal maze, will ensure the building's sustainable operation. Perching the building atop pillars, adding covered outdoor areas and incorporating setbacks in the facades could reduce the impact of the summer heat while the large building volumes store cool air, making for a more comfortable interior climate. A suitable insulation standard and a ventilation system with heat recovery ensure low heating and cooling loads and good air quality even given the school's high occupation density. The design responds to the prevailing climatic conditions in central Spain through a close interplay of architectural and energy concepts.

The school is more than a matrix of classrooms; it is the heart of the students' learning environment. It shapes both their understanding of the built and natural worlds and their sociocultural experience. The spatial compositions and visual connections within the school promote a sense of group identification and intercultural exchange, continuing the tradition and success of the German School.

1. 幼儿园
2. 小学
3. 初中
4. 自助餐厅
5. 礼堂
6. 体育馆

1. kindergarten
2. primary school
3. secondary school
4. cafeteria
5. auditorium
6. sports hall

二层 first floor

1. 幼儿园
2. 小学
3. 初中
4. 自助餐厅
5. 礼堂
6. 体育馆
7. 健身房
8. 门厅庭院
9. 前院

1. kindergarten
2. primary school
3. secondary school
4. cafeteria
5. auditorium
6. sports hall
7. gymnasium
8. foyer courtyard
9. forecourt

一层 ground floor

1. 幼儿园的热力空间
2. 小学的热力空间
3. 初中
4. 图书馆
5. 礼堂的热力空间+设备间
6. 体育馆设备间
7. 地下车库

1. thermal maze of kindergarten
2. thermal maze of primary school
3. secondary school
4. library
5. thermal maze + technique auditorium
6. technique sports hall
7. subterranean garage

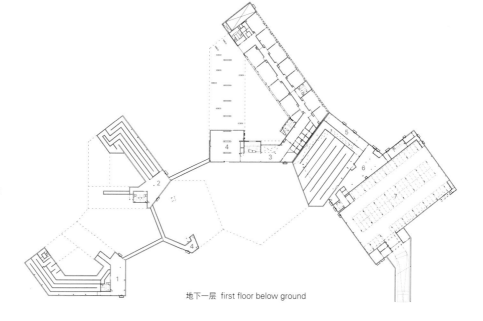

地下一层 first floor below ground

项目名称：German School Madrid / 地点：Calle Monasterio de Guadalupe 7, 28049 Madrid-Montecarmelo, Spain / 建筑师：Gruentuch Ernst Architects - Armand Grüntuch, Almut Grüntuch-Ernst / 项目管理：Bureau Veritas Construction Services / 项目团队：Erik Behrends, Benjamin Bühs, Florian Fels, Arno Löbbecke, Olaf Menk, Jens Schoppe, etc. / 结构工程师：GTB-Berlin Gesellschaft für Technik am Bau / 结构分析检测工程师：Mike Schlaich / 建筑设备：Ingenieurbüro für Haustechnik KEM / 能源技术：Transsolar Energietechnik / 能源概念竞赛：Klaus Daniels - HL-Technik / 防火：hhpberlin / 防火和建筑设备：Urculo Ingenieros / 开放空间规划：Lützow 7 / 照明设计：Lichtvision / 岩土工程：GuD Consult / 艺术：Carsten Nicolai, Folke Hanfeld / 客户：Federal Republic of Germany, represented by the BMUB, represented by the BBR; Association German School Madrid / 用地面积：34,700m² / 总建筑面积：27,065m² / 竞赛时间：2009 / 竣工时间：2015 / 摄影师：©Celia de Coca (courtesy of the architect) (except as noted)

P166 Chadwickdryerclarke Studio

Robin Dryer and Mark Clarke joined Chadwick Dryer Clarke as Directors in 2011. They have 20 years' experience in various areas. Robin graduated from the Newcastle University in 1997 and pursued several internal refurbishment projects in his early years. Moving to RMJM, Cambridge in 2003, his first project was the Film and Media Building at the University of Hertfordshire. After becoming Associate Director, he was also responsible for projects at the University of Bedfordshire and Coventry University. Mark has delivered a wide range of projects throughout the London and Cambridge area. His most notable being The Q building in Stratford, London, which was short listed for an RIBA award. Has worked in Cambridge for almost ten years, initially as an Associate for Frank Shaw Associates. He then moved to RMJM where he was responsible for the delivery of the Campus Centre at the University of Bedfordshire.

P74 SANNA

Was Established by Japanese architects, Kazuyo Sejima and Ryue Nishizawa in 1995. Kazuyo Sejima received M.Arch in 1981 from Japan Women's University and worked at Toyo Ito & Associates before founding Kazuyo Sejima & Associates in 1987. She has been teaching at Keio University, Politecnico of Milan, University of Applied Arts Vienna and Yokohama Graduate School of Architecture. Ryue Nishizawa received M.Arch in 1990 from Yokohama National University and joined Kazuyo Sejima & Associates in 1990. He was an Associate Professor at Yokohama National University and is currently teaching at Yokohama Graduate School of Architecture.

P148 Henning Larsen Architects

Is an international architecture company in Denmark, with strong Scandinavian roots. Founded by Henning Larsen in 1959, and is currently managed by CEO Mette Kynne Frandsen and Design Director Louis Becker. Has offices in Copenhagen, Oslo, Munich, Istanbul, Riyadh, the Faroe Islands and Hong Kong. Its goal is to create vibrant, sustainable buildings that reach beyond itself and become of durable value to the user and to the society and culture that they are built into. Its ideas are developed in close collaboration with the client, users and partners in order to achieve long-lasting buildings and reduced lifecycle costs. This value-based approach is the key to our designs of numerous building projects around the world-from complex master plans to successful architectural landmarks. Won the prestigious European Union Prize for Contemporary Architecture-Mies van der Rohe Award 2013. Kasper Kyndesen, Louis Becker, Jacob Kurek, Ingela Larsson, Werner Frosch, Mette Kynne Frandsen, Peer Teglgaard Jeppesen, Søren Øllgaard, and Lars Steffensen, from left. Signe Kongebro at the front.

P132 C.F. Møller Architects

Is one of Scandinavia's oldest and largest architectural practices. Was founded in 1924 by the now deceased Danish architect C. F. Møller. Today it has more than 350 employees in Head office in Aarhus, Denmark and some branches in Copenhagen, Aalborg, Oslo, Stockholm and London. Over the years, have won a large number of national and international competitions. Also, has been exhibited locally as well as internationally at places like at RIBA in London, the Venice Biennale, and the Danish Cultural Institute in Beijing. In 2015, it was named one of the Top 10 Most Innovative Companies in Architecture.

P196 Grüntuch Ernst Architects

Was founded in 1991, Berlin by Latvian architect, Armand Grüntuch(1963)[left] and German architect, Almut Grüntuch-Ernst(1966)[right]. Armand studied at RWTH Aachen and IUAV Venice. Almut studied at the University of Stuttgart and the AA in London. Both taught at the Hochschule der Künste, Berlin. Armand has been a member of the Advisory Board of the Bundesstiftung Baukultur since 2016. Almut has been the chair of the Institute of Design & Architectural Strategies at the Technische Universität Braunschweig since 2011. From 2010 to 2015 she was on the consultant committee for town planning in Munich, and since 2016, has been a member of the Academy of Arts Berlin.

P46 Guillermo Vázquez Consuegra

Is a Spanish architect, graduated from the ETSA Seville in 1972. Is Professor of Projects at this university until 1987. Was a Visiting Professor at the universities of Lausanne, Syracuse, Venice and Mendrisio, and Visiting Scholar at the Getty Center in Los Angeles. Is an Honorary Member of the American Institute of Architects since 2014. Has taken part in many exhibitions including 2004 Venice Biennales, The Museum of Modern Art, New York 2006, and IX BIAU Rosario, 2014. Received many awards including the Ibero-American Bienal Award 2014, the International Architecture Awards 2015 from The Chicago Athenaeum Museum, The Plan Award 2015. Recently has been awarded with the Spanish Architecture Gold Medal 2016.

P88 Ctrl G

Is an Colombian architecture studio founded by Catalina Patiño[left] and Viviana Peña[right] in 2008, Medellín. Has been a part of Colombian architectural collective called "Archipiélago". In 2015, they decided to disband Ctrl G to start two independent offices. Viviana founded Viviana Peña Taller de Arquitectura and Catalina founded Capa Arquitectura.

©Photopop

P182 CEBRA

Is a modern Danish architectural office, based in Aarhus, Denmark and in Abu Dhabi, the UAE. Is a member of the Danish Association of Architectural Firms. Was founded in 2001 by the architects Mikkel Frost[left], Carsten Primdahl[second-left] and Kolja Nielsen[third-left]. They are members of Danish Architects' Association and RIBA. In 2017, architect MAA Mikkel Hallundbæk Schlesinger[right] joined the group of partners. Most CEBRA projects are within the fields of education, culture and housing. Besides consolidating the CEBRA management with a new partner, the architects Lars Gemynthe Gylling, Thomas Bang Madsen and Flemming Svendsen have been appointed as Associates.

Francesco Zuddas

Is Senior Lecturer in Architecture at Anglia Ruskin University. Studied at the University of Cagliari and at the Architectural Association. Also taught at the AA, Central Saint Martins, and Leeds Beckett University. In 2013~14 he was a Visiting Research Scholar at GSAPP, Columbia University. Is currently writing a book based on his doctoral research on the project of universities in Italy in the 1970s, with the working title *The University as Settlement Principle* (Routledge, 2019).

Andrea Giannotti

Started working in Paris in 2004. His interest in multicultural and metropolitan environment led him to start working in diverse cities of Paris, Rome, Rotterdam, and Beijing. In 2009, he started collaborating with Archdaily and in 2010 with TheArchHive, writing articles and researching on modern and contemporary architecture in the urban metropolis, through a personal selection of projects.

P88 51-1

Is an architecture studio based in Peru (51), Lima (1) since 2005. Is one of the studios that made up collective Supersudaca (supersudaca.org) having bases -in addition to Peru- in Argentina, Chile, Uruguay, Curacao, Belgium and the Netherlands. As members of Supersudaca, cooperated actively in research, workshops, lectures and design projects. Was selected one of the 30 emergent studios from Iberoamerica by *2G Dossier* (Spain) in 2007. In 2009, *ICON* magazine considered them among 'the 20 architects that will change the future'. In 2013, they were selected for the III Latin American Architecture Biennale in Pamplona and in 2014 for the Iakov Chernikhov Prize. Cesar Becerra, Fernando Puente Arnao, and Manuel de Rivero, from left.

P102 **Shigeru Ban Architects Europe**
Was born in Tokyo, 1957. After studying architecture at the SCI-Arc and Cooper Union, founded his private practice in Tokyo, 1985. Has been a Consultant of United Nations High Commissioner for Refugees (UNHCR) and Jury of Pritzker Architecture Prize. Selected works include Nicolas G. Hayek Center, Centre Pompidou - Metz, and Oita Prefecture Art Museum. Is an honorary fellow of the AIA, RAIC, JIA and International Fellowship of the RIBA. Received an honorary doctorate from the Humane Letters Amherst College, Technical University of Munich and Cooper Union. Was a Visiting Professor of Architecture at Columbia University, Harvard GSD and Cornell University. Has been teaching at Kyoto University of Art and Design since 2011 and Keio University since 2015. Received multiple awards, including Grande Médaille d'or de l'Académie d'architecture (2004), Arnold W. Brunner Memorial Prize in Architecture (2005), Auguste Perret Prize (2011), Pritzker Architecture Prize (2014) and Mother Teresa Social Justice Award (2017).

P102 ©Shigeru Ban Architects Europe

P148 **GPP Arkitekter**
Is based in Denmark and work with a wide range of projects including public institutions, education, culture, sports, housing, commercial, as well as renovation. Lead by four partners, Carsten Gjørtz, Hans-Jørn Borgen Paulsen, Niels Haugard and Søren Madsen(CEO). Has won numerous competitions in Denmark and neighboring countries.

P62 **Allied Works Architecture**
Was founded in Portland, Oregon in 1994 by Brad Cloepfil. Opened the New York City office in 2003. Architect, educator, and principal, Cloepfil (1956) was born in Portland and earned his B.Arch at the University of Oregon. After working in the offices of Skidmore, Owings & Merrill in Los Angeles, and Mario Botta in Switzerland, Cloepfil moved to New York and received his Master of Science in Advanced Architectural Design from the Columbia University Graduate School of Architecture in 1985. Cloepfil has held guest professorships at Columbia University, University of California, Berkeley, Rice University, Cornell University, Syracuse University, and the University of Oregon. Guided by principles of craft and innovation, This 40-person practice creates designs that resonate with their specificity of place and purpose. Recent and notable works include the Clyfford Still Museum in Denver and the National Music Centre of Canada in Calgary.

P62 ©Adrian Gaut

© 2019 大连理工大学出版社

版权所有·侵权必究

图书在版编目(CIP)数据

学习中的城市：汉英对照 / 日本SANAA建筑事务所等编；贾子光，段梦桃译. — 大连：大连理工大学出版社，2019.4
（建筑立场系列丛书）
ISBN 978-7-5685-1956-4

Ⅰ. ①学… Ⅱ. ①日… ②贾… ③段… Ⅲ. ①学校—教育建筑—建筑设计—研究—汉、英 Ⅳ. ①TU244

中国版本图书馆CIP数据核字(2019)第067959号

出版发行：大连理工大学出版社
　　　　　（地址：大连市软件园路80号　邮编：116023）
印　　刷：上海锦良印刷厂有限公司
幅面尺寸：225mm×300mm
印　　张：13.25
出版时间：2019年4月第1版
印刷时间：2019年4月第1次印刷
出 版 人：金英伟
统　　筹：房　磊
责任编辑：张昕焱
封面设计：王志峰
责任校对：杨　丹
书　　号：978-7-5685-1956-4
定　　价：258.00元

发　行：0411-84708842
传　真：0411-84701466
E-mail：12282980@qq.com
URL：http://dutp.dlut.edu.cn

本书如有印装质量问题，请与我社发行部联系更换。

建筑立场系列丛书 01：
墙体设计
ISBN：978-7-5611-6353-5
定价：150.00元

建筑立场系列丛书 09：
墙体与外立面
ISBN：978-7-5611-6641-3
定价：180.00元

建筑立场系列丛书 17：
旧厂房的空间蜕变
ISBN：978-7-5611-7093-9
定价：180.00元

建筑立场系列丛书 25：
在城市中转换
ISBN：978-7-5611-7737-2
定价：228.00元

建筑立场系列丛书 33：
本土现代化
ISBN：978-7-5611-8380-9
定价：228.00元

建筑立场系列丛书 41：
都市与社区
ISBN：978-7-5611-9365-5
定价：228.00元